W9-CLF-663

Love Canal: Science, Politics, and People

Source: New York State Department of Transportation.

Aerial View of Love Canal Neighborhood

Love Canal: Science, Politics, and People

Adeline Gordon Levine
State University of New York
at Buffalo

LexingtonBooks
D.C. Heath and Company
Lexington, Massachusetts
Toronto

Library of Congress Cataloging in Publication Data

Levine, Adeline Gordon.
Love Canal.

1. Pollution—Toxicology—New York (State)—Niagara Falls.
2. Pollution—New York (State)—Niagara Falls. 3. Love Canal Chemical Waste Landfill (Niagara Falls, N.Y.) 4. Chemical plants—Waste disposal—Hygienic aspects—New York (State)—Niagara Falls. I. Title.
RA566.4N7L48 363.7′28 80-8361
ISBN 0-669-04034-7 Casebound AACR2
ISBN 0-669-05411-9 Paperbound

Fourth Printing, June 1983

Published simultaneously in Canada

Printed in the United States of America

Casebound International Standard Book Number: 0-669-04034-7

Paperbound International Standard Book Number: 0-669-05411-9

Library of Congress Catalog Card Number: 80-8361

To the people who lived at Love Canal,
who deserve the respect and admiration
of the world; for their struggle alerted us all
to the social and human costs of toxic waste
disposal.

Contents

Foreword

This book, an exemplary case study in classic sociological tradition, chronicles the efforts of citizens living in the Love Canal area of Niagara Falls to secure the resources they needed to escape homes poisoned by chemical wastes. It tells how they used tried-and-true methods of grassroots democracy in their knock-down drag-out struggle with corporate bureaucracies wielding more economic, social, and political power than the founding fathers of this republic ever envisioned. It tells how friends and neighbors, accustomed at best to sporadic citizenship activities, had to reorganize their lives and rearrange their priorities, dedicating themselves to nothing less than full-time citizenship—with lots of overtime.

You will read about their loss of innocence—how they had to shake the cobwebs out of their own minds to rid them of the Hollywood myth that collective problems can be solved by measures taken individually. To pinpoint the threat and dare to call it by its right name—life-threatening chemical poisoning—they had to demystify the legalese of bureaucrats and the professional jargon of scientists. And they learned to look with skepticism on the credentials of experts who did not breathe the air they breathed, drink the water they drank, and walk the streets of the Love Canal neighborhood they lived in.

Organizational skills had to be developed and leaders had to emerge to weld together the tight, feisty organization that converted indignation into goal-directed work and enabled them to move into the political arena and play political hardball with the pros.

Their grievances had to be entered into the public agenda and kept there as a pressing public issue. This meant learning how to reach the gatekeepers in pressrooms at newspapers, magazines, radio, and television—those men and women who have the power to move events stage front and center, or keep them out of the public's line of sight, immune from its judgments.

It took three years for all gears to mesh, but finally hundreds of families received the funds they needed to get out of their poisoned homes and make a new start.

Part of the Love Canal story is a success story: government *can* be forced to accountability when citizens pool their strengths and use publicity and votes as bargaining chips in the game of power politics. But the story is, at the same time, sobering testimony to the eroded meaning of a phrase like *public accountability* in our modern, high-technology society.

The people of Love Canal can walk away from their homes, but it is not at all certain that they can walk away from the chemical contaminants they lived with for so many years. Are those poisons still stored in blood and bone, tissue and germ cell? Are they incrementally distorting developmental life processes in ways that will become measurable only in twenty or thirty years, or turn up as diminished life chances for the next generation? Does the dollar indemnity that government pays today close the books on public accountability for outcomes that become visible only tomorrow?

And what about the accountability of the Hooker Company and its parent corporation, Occidental Petroleum? Their role in the events at Love Canal highlights the way our bizarre rules of social bookkeeping relieve business of much of the cost of doing business. A company that produces chemical compounds for private profit unloads into the public environment chemical wastes that are as intrinsic a part of their production process as the trademarked products the company sells. Yet the cleanup costs are passed on to a public that pays in two ways: funds paid out by government and human costs paid out by the people of Love Canal.

This book mentions the low profile the Hooker Company has maintained throughout the controversy. This is still their policy. Management has shown remarkable consistency in hewing to the company's public-relations line: soft-pedalling public comment, avoiding specific issues, insisting the whole affair has been exaggerated, and disclaiming responsibility. Such solidarity attests to the ideological power of the structural constraints built into all corporate bureaucracies on the scale of Hooker. They have been well analyzed by Christopher Stone, (*Where the Law Ends,* 1975). Decisions are made by unnamed personnel, some of whom are long since gone. Knowledge is fragmented and responsibility diffused through a long chain of command. Evidence is kept off the record or hidden in an avalanche of memos buried in dead storage. Even when responsibility can be pinned down, organization men with the same interests, loyalties, and reward systems close ranks and keep quiet; for he who points the finger breaks the code and is punished. Through it all, the depersonalized, rationalized, narrowly instrumental social structure of corporate bureaucracy shields the rest from having to face up to their role in shaving the points of moral scorekeeping.

Professor Levine shows how these same processes play themselves out in the corporate structures of the public bureaucracies citizens turn to for redress of grievances. She focuses the spotlight especially on the doctors and scientists in bureaucratic employ or in a client relationship with the bureaucrats. Matters requiring social judgment are rephrased as technical questions in the exclusive domain of their expertise. But like all questions, theirs, too, have built-in assumptions, and objective scientific facts must still be rearranged and interpreted. Judgment is called for more than calculation, wisdom more than the smarts. The book makes it abundantly clear that the most highly respected credentials do not lessen by one whit the power of bureaucratic social structures to encourage the narrowest definitions of self-interest; nor do they reduce the age-old human tendency to allow judgment to be colored by loyalties and position in the pecking order.

This book can be read as one more addendum to Rachael Carson's 1962 documentation of the chemical industry's insults to the environment and the tendency of watchdog agencies to look away; as a muckraking revelation of what goes on in high places; as a history of events in the lives of groups of Americans who faced severe threat with extraordinary courage, ingenuity, and inventiveness. But it is also a study of the sociology of two contrasting social structures that differ in scale and in their mode of organizing the way people relate to each other. One is the techno-corporate structure of bureaucracy, which mechanizes human interaction just as commodity production is mechanized. The other is the structure of groups like the Love Canal Homeowners Association. Techno-corporate structures distance people from each other; the association brings people together. In techno-corporate structures it is difficult to read each other's faces, walk in each other's shoes, and respect each other's feelings as we respect our own; in the association it is difficult not to. Techno-corporate structures fragment work and disguise the outcomes of one's daily efforts; the association requires people to act together; their tasks are out in the open, decisions are shared, and democracy becomes a daily, lived experience. Techno-corporate structures breed *apparatchiks* who act according to a rule book that has no entry for the word conscience. The association runs by rules but no rule book, and conscience is a potent ally.

In short, structures like the association make it easy for people to live up to their best potential, and provide maximum opportunity for its daily exercise. This has been known to happen in techno-corporate structures, too, but mainly among the few hardy souls who dare to buck the system.

At thousands of dumpsites in this country alone, chemical wastes are still seeping into topsoil, subsoil, waterways, and aquifers, and more are dumped every day. Cleanup efforts are still delayed as disputes go on in the same terms of discussion the book details: who pays, how much, and who decides? The bottom line is still calculated in terms of how clean a cleanup we can afford. The Love Canal story reminds us that wise social policy must look to a different set of accounting principles. Can our planet afford an industrial system that forces her to digest ever-increasing doses of nonbiodegradable junk? Can our democracy afford corporate bureaucracies in business and government that wield so much organized power that ordinary Americans who dare to buck them must—like the Love Canal residents—give up years of their lives to the struggle?

Rose K. Goldsen
Cornell University

Acknowledgments

I want to thank all the people who lived in the Love Canal neighborhood; who worked there; who were concerned or involved in any way with the Love Canal crisis; and who generously shared their information, experiences, and impressions with me. The list of those to whom I am indebted is so long that I cannot name them without slighting others equally important in the development of this book.

I also want to acknowledge the hard work, dedication, and enthusiasm of the sociology graduate students who worked and studied at Love Canal: David Barker, Martha Cornwell, Susan Hufsmith, Sharon Masters, and Penelope Ploughman during the first year; and Gloria Brennan, Phil Gray, and Pat Tirone during the second year. Those who have experienced the intellectual excitement of working in a new area of research can understand how important their contributions have been to my thinking.

The Research Foundation of the State University of New York and the Moir Tanner Award administered by the University of Buffalo Foundation provided much-appreciated funds to help reimburse day-to-day research expenses during the first year of the project.

I also want to express my thanks to the many colleagues who talked with me as the work progressed, read various portions of early drafts, shared their expertise, and encouraged my efforts by their interest. I am particularly grateful to Murray, David, and Zach Levine, and to Susan Shackman for being there with their encouragement and criticisms. And finally, my thanks go to Nancy Holdsworth for her consistently cheerful, helpful responses to my requests for assistance.

Love Canal: Science, Politics, and People

1 Introduction

Background

A new type of disaster emerged at Love Canal—a result of the indiscriminate use and disposal of synthetic products, leading to physical harm to people and environment and severe disruptions of the social fabric. At Love Canal, chemical leachates had oozed from a long-forgotten chemical-waste-disposal site into a residential area of Niagara Falls, New York. (Leachates are the liquids resulting when water percolates through a chemical landfill.) The complex drama that resulted captured the attention of the whole world, for what happened at Love Canal not only affected the lives of those directly involved but foreshadowed issues that face us all. In the United States alone, there are hundreds, perhaps thousands, of chemical-waste-disposal sites similar to Love Canal.

The Love Canal disaster developed over several decades, as the impact of leaching chemicals is uncertain and slow in developing. The visible effects are limited, and some may be attributed to other causes. When today's uninformed visitor to the Love Canal neighborhood feels a chill, it is from the sight of abandoned homes, with boarded up windows and doors and overgrown yards, all surrounded by a high chain-link fence. What the visitor sees are signs of the reactions to things that lie essentially hidden, detectable only by special procedures. Chemicals are present under the ground, in the yards and houses, and in the air, in people's bodies; but Love Canal is a conceptual event, for the physical manifestations can be readily overlooked, ignored, denied, and minimized.

The Love Canal story started many years ago, but everything came together in the summer of 1978. In a sense, there was no crisis or disaster until authorities defined it publicly and the event was reported in the world press. Unlike more familiar disasters, with known and uncontrollable moments of impact, the Love Canal emergency could have been defined earlier or later. The social actors at Love Canal were affected and behaved according to their perceptions and interpretations of the situation. This book is the story of the behavior and attitudes of various actors and the social forces that inhibited or encouraged their changing definitions of the situation and thus their response to the Love Canal events. It is also a story about choices and values, and it provides an illustration of the complex ethical dilemmas we face in balancing human health and well-being against economic costs and benefits.

Both individuals and institutions reacted to Love Canal with slowly dawning awareness of the far-reaching consequences of past activities: burying chemical wastes and encouraging people to live in homes and send their children to school close to the disposal site. The consequences might or might not have been anticipated by the people who decided long ago to bury and to build, or who allowed these things to be done. Their decisions and actions affected the future, but the individual decision makers were not accountable at the time of their actions. The day of reckoning was deferrable, with the impact and responsibility to be borne by someone else in the distant future.

The consequences of actions taken at Love Canal, though slowly revealed, have been of a greater magnitude than was expected when the situation was first publicized in the summer of 1978. Because there are unknown numbers of similar hidden disasters, the methods for the mitigation of present problems and the prevention of others in the future have been very much part of the considerations of concerned corporate, governmental, and citizen groups. Every decision at Love Canal had the potential for setting precedents. Issues of power in the control and allocation of resources, of who takes the risks and who benefits, of the use of science in the service of power, of pure political maneuvering, and of whose story reaches the public have been important factors throughout the story of Love Canal. This book describes these processes and the complexities of developing solutions to new problems in American society.

Research Approach

I first heard of Love Canal on a local television news broadcast in early August 1978. I saw people at a meeting of some sort, shouting and crying, and a number of young children, some wearing crudely lettered signs proclaiming that they did not want to die this year.

A few days later a colleague and I drove up and down 97th and 99th Streets in Niagara Falls.[a] We saw compact homes with neat lawns, carefully tended gardens, and shade trees cooling the streets. It seemed incongruous to see plywood panels over some doors and windows, moving vans drawn up to curbs, and men and women carrying boxes and bundles out to their cars. I felt an eerie sensation that danger was lurking somewhere in this peaceful setting, and I became intrigued with the problem. After a long day spent talking with New York state personnel in hastily organized offices, with residents who were unpacking in temporary quarters, with the president of the newly formed residents' association, and with other people as

[a]Thanks are due to Barbara Howe, Department of Sociology, SUNY/Buffalo for that first visit.

well, we returned home emotionally wrung out and exhausted by the effort of making some sense of what we had just witnessed. By the next morning, I was already "hooked." Love Canal soon became my continued obsession, and the complex story continues to unfold.

In the days that followed, my immediate question was what sort of research I would do, for it was obvious that there were a number of events to examine from a sociological viewpoint. It takes resources to do any sort of research, of course, and I attempted to deal with that matter first.

After several visits, telephone calls, and written inquiries to various foundations and agencies, I concluded that little money was available for pursuing research on short notice, in new areas, and in an exploratory way. The few agencies expressing interest required extensive detailed proposals. Therefore, rather than spending my days applying for grants, I chose to devote my time and energy to the swiftly moving and changing events.

In order to set to work immediately, I decided to use the resources readily available to me at the State University of New York at Buffalo. I organized a field-research seminar on Love Canal with five advanced graduate students, who responded to my invitation with enthusiasm. Although we initially planned to study the Love Canal families' responses to stress, we quickly realized that there was a great deal more to be learned, and we decided to find out everything we could.

One person was most interested in the state task-force organization, another in older people and minority groups, another in the largest citizens' organization, one was concerned with issues of family stress, one with general corporate and governmental responses, one with the role of the mass media. We prepared interview schedules based on our knowledge of the situation, interviewed families, visited the Love Canal area frequently and regularly, and attended all public meetings and other events, so that no week passed without extensive contacts between the research group members and the Love Canal people and activities.

In the first year, the students and I acted as controls for each other, checking facts and questioning the inferences that could be drawn from our observations and interview data. Several key participants met with our seminar to discuss their experiences. Although we came to understand best the views of the groups we were with in the field, we tried to keep our minds open to the various views of the different groups. We thought of our method of presenting and advocating opposing views as an "adversarial discussion method."

In the second year, new students joined me for a semester. We discussed the possibility of further research projects, and they, too, visited Love Canal and its people regularly. During the final year and a half, although I remained in contact with the students, I worked alone, discussing my ideas with colleagues, participants, students, and lay and professional audiences.

I continued to visit Love Canal at least weekly, talking with people at the Love Canal Homeowners Association office, talking with others by telephone, attending public and private meetings, and interviewing others as events required. I continued to collect every available document from the stream of reports, newsletters, and other information, in addition to newspapers and other published material.

This book is based on all these experiences and data. The data include formal interviews with 102 people from 61 families; reinterviews with 63 people from 38 families; formal interviews with 30 state task force members; individual interviews with 20 other key people; almost three years of field notes; and a collection of reports, letters, other documents, photographs, magazine articles, and several thousand newspaper items. Much of the information was distributed publicly at the time, was freely shared by participants, or, in a few instances, was obtained through requests under Freedom of Information Laws.

Newspaper reporters were on the scene day after day, week after week. I found that their accounts were usually highly accurate, and, since I was close to the events, I could check the occasional discrepancies. I realized that the newspaper articles provided detailed documentation of key events by skilled observers, producing an informal current history for my use.

My university base provided immediate access to the crucial resources of ideas and advice of colleagues in several disciplines. An anthropologist joined us when we considered problems of bias in field research, and we had the advice of a social psychologist, a clinical-community psychologist, a lawyer, and others as we needed them. Thus, it was possible to do a great deal of research on an important topic, using the resources of a university without outside support.

The creativity, knowledge, time, and energy of the graduate students who worked with me were invaluable. They benefited, too, from a wealth of real field experiences and professional participation they could have obtained in no other way. My academic base was critical, as I worked in a social setting where there is an expectation that one does research. Most important, I had the freedom to study and to describe whatever I thought important and appropriate for a sociologist to pursue, using my professional judgement. The observations and opinions of observers who are independent of official governmental or corporate efforts have been very important at Love Canal. I consider myself in that independent category and thus am able to act in the academic's traditional role of social critic.

I have been concerned about bias in my findings and presentation, for that is always a problem in research. I have tried to find out as much as possible about all sides and to understand the perspectives of people in various positions. My greatest amount of contact was with the residents, however, and they offered me a wealth of information, virtually opening their lives to me. They wanted the world to know everything, whereas others were sometimes happier to keep some matters obscure. The difference too was that the residents were fighting for survival.

Although I am critical of some governmental actions at Love Canal, in no way should this book be considered an argument against government intervention. Rather, it is a plea to make those essential protections work better than they did at Love Canal.

I want to make a final comment as a sociologist. As I worked to make sense of the experiences of Love Canal, I used standard sociological conceptions. We do have useful intellectual tools, especially when the sociological imagination helps to show us how individual troubles and individual behaviors reveal the larger social world.

2 The Beginning and Before the Beginning

"A review of all the available evidence respecting the Love Canal Chemical Waste Landfill site has convinced me of the existence of a great and imminent peril to the health of the general public residing at or near the site" (Whalen 8/2/78*a*). Those solemn words, read aloud by Dr. Robert Whalen, New York State's commissioner of health, on August 2, 1978, were part of an official order that confirmed the worst fears of some residents of Niagara Falls, New York, and was a conceptual bombshell for thousands more in the Love Canal region. The order was intended to solve the manifestations of a new public-health problem, but solutions are still elusive after years of turmoil for a thousand families and after the expenditure of millions of dollars of public money. The order made the nation and the world aware of the hazards of abandoned toxic-waste-disposal sites. It set into motion events that occupied the thousand families for the next two years, and more, and will influence the course of their lives henceforth. It initiated events that involved every level of government and laid bare the complexities of developing social solutions to newly defined public problems within a political context. The events also demonstrated the limited contribution of scientific knowledge to political decision making.

The official order declared that, in a pleasant residential area of the city of Niagara Falls, materials hazardous to human life were leaching from an abandoned chemical-waste-disposal site located in an old canal. More than two decades earlier, an elementary school had been built over the filled-in canal, and the school became the center of a residential neighborhood. It was hard to believe, but true, that chemical leachates had been moving out from under the grassy fields and playgrounds of the school, through the grounds and neat lawns, and right into the basements of the homes that lined the shady streets.

Before the Beginning

The history of Love Canal extends to the geological ages when the Niagara River was first formed and began to flow north, connecting Lake Erie with Lake Ontario, some thirty thousand years ago. The Niagara waterfall was a freak of nature, carved out of complex rock formations, forcing the river over two crests to thunder down, hundreds of thousands of gallons per second,

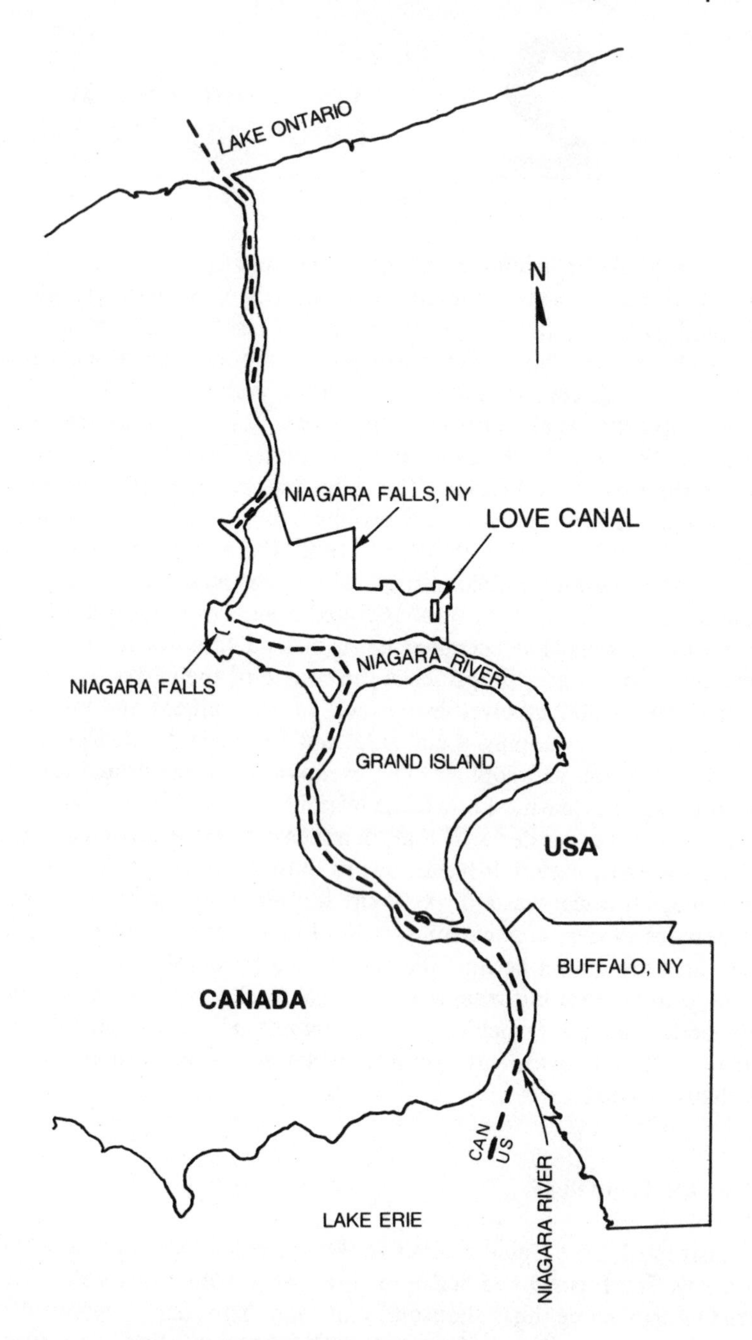

Source: Map was drawn by CWP Graphics, Washington, D.C.

Figure 2-1. The Location of Love Canal

almost two hundred feet to the Niagara gorge below, midway on the river's journey between the two Great Lakes.

The swiftly flowing river's potential for generating power was noted by the region's early explorers and settlers, one of whom built a tiny canal for a sawmill in 1757. In the 1830s, an engineer described the river's great potential as a source of cheap water power. In 1879, hydroelectric power was used to light up fountains in a nearby park for the thousands of tourists who came, even then, to view the falls by day and night. Then, in the 1890s, William T. Love made a power-producing canal the centerpiece of his plan to attract industry, through free energy, to a planned community, Model City, to be built on the shores of Lake Ontario. The canal was to run north from the Niagara River, capturing the mighty force of the water as it sped into rapids before rushing over the huge drop. The canal was to terminate a few miles farther north, flowing into the Niagara well beyond the waterfall.

Love's powers of persuasion and remarkable political connections helped him secure legislative permission to charter a development company, condemn properties, and divert as much water as necessary for the project. Model City was beyond the planning stage, with a factory and a few homes built and the ground broken for the canal at both ends, when things began to go wrong. When the depression of the mid-1890s struck, investors withdrew money from the company. Louis Tesla's discovery of a way to transmit electrical power by alternating current made locations near the source of water-powered electrical energy less essential for manufacturing. Congress passed an environmental bill, restraining the removal of water from the Niagara in order to preserve the waterfall. So Mr. Love's canal was left a monument whose name was forgotten for years. Estimated to be some sixty feet wide, ten feet deep, and three thousand feet long, the canal was embedded in an area of orchards and farms, watered by riverlets and creeks stemming from the Niagara, in the pastoral village of LaSalle, to the east of the city of Niagara Falls (Parry 1975).

Although Love's project failed, the development of the electrical industry attracted the chemical industry to the nearby area, because abundant, cheap electrical power is an essential element in chemical production. The Hooker Electrochemical Company was started in Niagara Falls in 1905 by Elon Hooker, employing seventy-five people in the hazardous tasks of manufacturing chlorine and caustic soda. An important part of the community from its earliest days, the company grew steadily and spurted ahead markedly after 1954, going from $19 million in sales in 1945 to $75 million in 1955 (Thomas 1955). In 1978, Hooker, now a division of Occidental Petroleum Corporation, employed 18,000 people worldwide and had net sales of $1.7 billion. The plant at Niagara Falls is the largest of sixty Hooker manufacturing operations; corporate headquarters are also housed in that city (Occidental 1978). The chemical industry is important in New York State and predominates in Niagara Falls and the surrounding area, with other nationally known chemical firms (Carborundum, Olin-Mathieson, Union-Carbide, and DuPont) established in addition to Hooker.

In May 1927, Niagara Falls annexed the village of LaSalle, making it and Love Canal part of the city. The canal was centered in a piece of land about two hundred feet wide. The area ran north from the Niagara, starting about one-quarter of a mile from the river's shore, with a drainage trench leading from the canal into the river. Along with nearby streams, small rivers, and ponds, the canal provided a pleasant place for outdoor play, swimming, and fishing.

In 1942, the Niagara Power and Development Company, successor to the Niagara County Irrigation and Water Supply Development Company, gave Hooker permission to dispose of wastes in the canal and then, in 1947, sold Hooker the sixty-foot-wide canal and two seventy-foot-wide strips of land abutting it to east and west. These innocuous business transactions were the precursors to the problems that would envelop the area thirty years hence.

Hooker used the site for waste disposal between 1942 and 1952 (Cull 11/17/78). They disposed of more than 21,000 tons of various chemical wastes at the site, at a depth of twenty to twenty-five feet (T.R. Johnston 7/13/78; USHR 3/21/79, Davis testimony, pp. 488-494; Hammersley 5/20/80). Many of the substances were known to be dangerous—caustics, alkalis, fatty acids, and chlorinated hydrocarbons from the manufacture of dyes, perfumes, solvents for rubber and synthetic resins, and other products—and were buried in metal or fiber barrels or in the form of sludge or liquid (Cull 11/17/78). At that time, and until the mid-1970s, waste disposal was a relatively simple matter. A would-be dumper simply acquired a permit from the local health department, which ascertained that the site would not endanger public health by attracting flies and vermin—a solution to a problem very different from that posed by chemical-waste disposal (Friedman 12/1/78). Hooker company officials considered the old canal an excellent dump site; it was large, lined with walls of thick, impermeable clay, and located in a thinly populated area where zoning regulations did not prevent waste disposal.

Old-time Love Canal area residents remember well, however, when municipal dumping began to foul their swimming and fishing spot, and then when the old canal began to serve as a disposal site for chemical wastes. Vivid testimony from long-time residents depicted the odious nature of the materials buried in Love Canal. In a letter to the *Niagara Gazette*, a woman said that, in 1943,

> they dumped all year early in the morning. . . . my husband was home sick and the smell was so bad that we could hardly breathe and we had to put

> wet towels over my husband's mouth and nose. It came over toward the houses like a white cloud and killed the grass and trees and burnt the paint off the back of the houses and made the other houses all black. I know for sure they did it all in '43 because my husband . . . died October 28, 1943 with lung trouble and I think the gases that came this way helped shorten his life. (A. Warner 7/26/78)

Karen Schroeder, a resident from childhood, recalled that, when she was seven years old, in 1952, the Love Canal disposal site was still being filled behind her family's home on 99th Street. A reporter described it:

> Workers would run screaming into her yard when some of the toxic chemicals they were dumping would spill on their skin or clothes. She remembers her mother washing them down with a garden hose until first aid could arrive. (Pollak 6/5/77)

Her mother said that, when the dumping of barrels was going on, "If a drum containing a certain material would break, the air would hit it and it would catch on fire. It seemed Hooker was always out here putting out fires" (Pollak 6/5/77).

And so it had started: the company dumping wastes; the residents suffering from their proximity to the chemicals—though they did not yet fully realize it; the regulations obsolete; and the regulators uninterested.

By 1953 the old Love Canal was nearly full. After a time, the site was covered over with earth or clay and grass and weeds grew on its rough surface, so that, many years later, it looked like long, broad fields. The first portend of the chemical disaster to come appeared in a disclaimer in a legal document, deeding the site to the Niagara Falls school board.

For the token payment of one dollar, the school board of Niagara Falls purchased sixteen acres of land, including the almost-filled canal (Niagara Gazette 10/17/53). The Hooker Company retained the right to use some portions of the land until they could buy other property for a waste-disposal site (Boniello 5/5/53). The deed of sale for this bargain, signed on April 28, 1953, by the executive vice-president of Hooker, included a disclaimer that read, in part:

> Prior to the delivery of this instrument of conveyance, the grantee herein has been advised by the grantor that the premises above described have been filled, in whole or in part, to the present grade level thereof with waste products resulting from the manufacturing of chemicals by the grantor at its plant in the City of Niagara Falls, New York, and the grantee assumes all risk and liability incident to the use thereof. . . . as a part of the consideration . . . thereof, no claim, suit, action or demand . . . shall ever be made . . . against [Hooker] . . . for injury to a person or persons, including death resulting therefrom, or loss of or damage to property caused by, or in connection with or by reason of the presence of said industrial wastes. (Deed recorded July 6, 1953, County of Niagara)

The disclaimer was not enough to dissuade the Niagara Falls school board from proceeding with their plans to acquire land, at no cost, in an area of burgeoning population; neither was the statement from the city's deputy corporation counsel, explaining that the board would have placed on it the risk and possible liability (Boniello, 5/5/53; Westmoore, 8/9/78); nor was the advice they received from the construction firm they hired to build an elementary school on the site. After construction work began, in January 1954, the school's architect wrote to the chairman of the board's education committee and told him that, when the contractor began to excavate, he reached a "pit filled with chemicals" and learned that there were two dump sites filled with chemical waste in fifty-five-gallon drums. The architect advised that it was "a poor policy" to build on the canal soil because of the odors and that, since he did not know what sorts of materials were in the soil, he would assume they might injure the concrete foundations of the school (Thiele 1/21/54). Faced with these difficulties, the school board ordered that the building be sited eighty-five feet north of the original location (Thiele 1/29/54). The kindergarten play area was moved from the place originally planned for it when it was discovered that a chemical dump lay beneath that portion of the ground (Thiele 10/15/56, 6/25/56). A drainage system was installed, apparently to correct a serious problem of water accumulation in the vicinity of the school. The drainage system connected to the storm sewer system and thence into the Niagara River (Thiele 10/15/56, 10/18/56; F.J. Lang 10/23/56; Curts 11/23/56; Packard 10/19/61; Palazzetti 8/27/78; MacClennan 5/11/79).

Only in later years would the gravity of all those decisions become apparent. At the time, work proceeded on the 99th Street Elementary School. After its completion early in 1955, 400 children attended it each day, playing at recess and after school hours on the playground and the adjoining fields, all over the old canal site.

At present, the issue of responsibility for damages is in the courts. Hooker asserts that they warned the school board at the time of purchase through the disclaimer in the deed and again in 1957, two years after school was completed, when Hooker's attorney warned that the land was suitable only for a school or for recreational purposes. In 1957, the company's attorney pointed out that damage to water and sewer lines and building foundations as well as personal injuries could result from the residues buried in the property. Why land inappropriate for other structures was considered suitable for a school is not clear. There was evidently no attempt by the company to warn the general public of the potential problems in 1952, when the school board, according to Hooker's attorney, intended to acquire the land by condemnation proceedings (Hooker 1980; A. Wilcox 11/21/57).

The action on both sides is puzzling. The school and its playground occupied less than five acres of the site. The costs for that amount of land in that part of the city must have been trivial in relation to the costs of constructing a modern school building, complete with auditorium and gymnasium. The company's concern about condemnation proceedings is also puzzling, since fair market value is paid for private land when it is taken for public purposes. One can only speculate that the gift to the school board may have resulted in some tax benefit for the company.[1]

As the years went by after the school was built, modest two- and three-bedroom homes went up, with backyards bordering the lands extending from both sides of the canal. When the last group of new homes was sold in the early 1970s, some were guaranteed by low-cost FHA mortgages. Some buyers had used GI mortgages. The homes, purchased for the most part by blue-collar workers, were good buys, with prices ranging from $18,000 to $23,000 in the early 1970s. Since the land the homes were on was not part of the Love Canal transaction between the school board and Hooker, the property owners' deeds carried no statements of disclaimer to give them written notice of what lay buried beneath the schoolyard and the adjacent land. Evidently none of the residents considered that they were in a hazardous location.

There were no provisions for monitoring or keeping intact the huge underground chemical container in the midst of the homes. Streets were put in, paralleling the canal site, and more homes were built after the school went up. Two roads were cut through land from east to west, framing the school and its playground, and sanitary and storm sewers and utility lines were installed for the school and the homes nearby.[2] All the underground construction necessary for a modern residential neighborhood cut channels through the supposedly impermeable clay walls lining the hidden canal. Soil was moved from the canal surface at various times for use as fill or topsoil when homes were built, a practice also followed by residents from time to time. The original cover of the canal—clay or dirt—began to age, cracking and chipping over the years. In the 1960s, an expressway was built across the southern end of the original drainageways leading from the canal, thus possible cutting off the flow of leachates from the canal into the Niagara River.

By the mid-1970s, only a few old-timers remembered the whole history of the canal that had turned into a schoolyard and open fields. Most of the people there had arrived in the mid-1960s and later, after the school and road construction was complete, and were unaware of the early history of the place. There were signs, however, that possibly could have alerted the residents. The surfaces of the large fields north and south of the school's playground were rather uneven, for example, and from time to time holes would open up when barrels deteriorated and the soil settled (Hart 7/28/78). People who noticed did not realize that such events were indications of something that eventually would reach out to affect their lives.

Many families either paid no particular attention to the canal's surface or assumed that someday it would be "all fixed up and nice"; when they bought their homes, they were told by real-estate salesmen and also read of cooperative plans between the Junior Chamber of Commerce and the city for a recreational park in the empty areas to the north and south of the school (*Niagara Gazette* 11/24/59; 7/14/60; 10/17/53).

From the early 1950s, there were times when strong odors were very troublesome, particularly after rainfalls or in humid weather, but these were usually attributed to the chemical odors from the busy plants a few miles away in town—a fact of life for residents and employees in a town full of chemical plants. Sometimes children's feet became irritated if they played barefoot near the school, and dogs' noses and bellies grew red and irritated after they ran on the big field. Parents washed the children, put lotions on their feet, and told them to be more careful. Residents recall, also, that rocks from the fields would often explode if they were dropped or thrown, but few complaints were made to city or county officials. Most people did not make the connection between the locale and the injuries; those who did thought at the time that the effects were too mild to cause concern.

Occasionally, incidents were related to the Hooker Company by worried residents, especially when materials appeared on the surface of the canal, and there are records of a case Hooker employees investigated in 1958 when three or four children were burned by "material at the old Love Canal property" (USHR 3/21/79, Wilkenfeld testimony, p. 615). There are no records to date, however, that anyone warned the public that the surface of the huge disposal site might be a dangerous place for children to spend time.

A man employed on the crew installing streets across the canal site said, "When they started putting that street through, that's when I became aware of what was in there because the fumes would make your skin real itchy and irritated and break out in all little blisters." There were evidently work stoppages when drums were exposed and oily liquids and fumes were released. (Hart 7/28/78). Like many others, however, who felt safe because "*they* built a school there" or "*they* gave us a 235 [FHA] mortgage," this man did not complain to any official body. It simply did not occur to him to do so. When people did complain, either the callers were ignored or short-term remedial measures were taken.

In 1965, one resident began to express concern about the black, oily substance in his basement sump pump. He complained to City Hall repeatedly, for he had to replace the pump with increasing frequency as time went on. By 1969, city records indicated a number of complaints and showed that city building inspectors had examined an area at or near the Love Canal dumpsite. The inspectors reported that the conditions were hazardous with holes in the surface of the field, formed by rusting barrels collapsing, and chemical residues left on the surface after rainwater had evaporated (Pollak, 10/3/76). Still the pattern of ignoring clues persisted.

Things began to change for the worse in the mid-1970s. For several years, the Buffalo-Niagara Falls area suffered a great increase in snow and rainfall. Love Canal residents noticed that, after the heavy rains, chemical odors were increasingly troublesome. Greatly increased flows of water seeping into the canal through the cracked surface and underground waterways may have accumulated faster than they could drain out; decades-old chemical barrels were deteriorating; chemicals were moving through underground waterways and through the soil to the yards and basements and vaporizing into the air in people's homes and yards. Whatever the path or the cause, more residents became suspicious that something was wrong.

Huge puddles appeared in some backyards, remaining almost permanently or taking days to evaporate after rains. Thick, black material appeared on some basement walls and floors. Some people found their backyards increasingly unpleasant and unusable. By the mid-1970s the central areas of the big fields were barren or supported only scrubby grasses. Residents recall that, during the spring of 1975, a hole appeared in the baseball diamond, near one of the bases, then another opened up a year later on the playground. At one point, the Niagara Falls school superintendent sent a message to parents asking them to instruct their children to stay on the sidewalks near the school (*Niagara Gazette* 11/4/76). Once the hole was filled in, that restriction was lifted; the symptom had been covered, literally.

In 1976, however, a fresh perspective led to a quietly launched investigation, undertaken not because of the residents' distress but for a different reason, and by an unexpected source. After that, events began to move—slowly at first—as various people who could do something became aware of problems at Love Canal or acknowledged that they existed or might exist.

1976 to 1978

The first source of the investigation of Love Canal was the International Joint Commission, which represents the Canadian and U.S. governments in monitoring the conditions of the Great Lakes (U.S. Congress 1911). Early in 1976, the commission detected traces of the insecticide Mirex in Lake Ontario fish. Soon afterward, the New York State health commissioner issued a warning that Lake Ontario fish were unsafe for consumption. The Department of Environmental Conservation for New York (DEC) launched an investigation, and one source of the Mirex was traced to Niagara Falls and to the 102nd Street dump site, next to Love Canal (Friedman 12/1/78; Swan 1979).

The dump was used both by the Hooker Corporation, after the Love Canal was filled, and by the Olin-Mathieson Company. When William Friedman, regional director of DEC, questioned officials of the two companies, both denied putting Mirex in that site—Olin's representative claiming that they had placed no toxic chemicals there and Hooker's stating that they had not placed Mirex in the spot. Friedman requested an accounting from the Hooker official of exactly what the company had disposed of and set a due date of mid-November 1976 for the accounting. (Friedman, 12/1/78; USHR 3/21/79, p. 646; Russell 11/9/76). Two years later, in December 1978, the accounting had not yet arrived, although Friedman had been told that a letter was sent in November 1977. Friedman had poor returns from another quarter as well. He had submitted samples of materials taken from the dump and from the Hooker "outfall" to the New York State Department of Health (DOH) laboratories, which perform analyses for the DEC. By December 1978, he had not received the analyses and had been informed that the samples had been lost, probably when the DOH had moved to new offices (Friedman 12/1/78; Pollak 5/1/77).

Other interested investigators were to have better luck however. Early in October 1976, a *Niagara Gazette* front-page story traced the history of the Love Canal landfill, describing its use as a chemical-disposal site, the sale of the land to the school board, and the history of the previously sporadic but now increasing complaints from residents (Pollak 10/3/76). This article suggested the danger for people living nearby. Shortly afterward, the journalist, David Pollak, went to the home of the man who had been complaining since 1965 about his sump pumps wearing out as they labored over black, oily substances in his cellar. Pollak scooped up a sample of the sludge and asked a friend at Chem-Trol, a large waste-treatment company, to analyze it, pledging not to reveal the analyst's name. Shortly thereafter, the journalist publicized the fact that the sludge sample definitely came from one of Chem-Trol's chief clients, the Hooker Chemical and Plastics Corporation (Russell and Pollak 11/2/76).

The ongoing DEC investigation and the chemical discoveries were the subject of a series of news articles in the late fall of 1976 and early winter of 1977 (for example, Russell and Pollak 11/2/76, 11/4/76; Russell 11/6/76, 11/9/76; Horrigan 11/4/76). A senior public-health engineer for the county health department commented, according to a news article, that "it was far better that the material be isolated at the places people live than continue to be pumped directly into the [Niagara] river." He also was of the opinion that, since the toxic materials were entering people's homes, the homeowners were responsible for ending the discharges at that point. He thought it would be a good idea for people to cement over the drainage holes in their basements. "Everyone has to do his part to clean up the environment and I don't think this is asking too much," he said (Russell and Pollak 11/4/76).

Friedman evidently did not share that public officer's sanguine attitude. In December 1976, he was angry when the representatives from the Hooker Corporation once again denied responsibility for the problem. The land was no longer theirs, they argued, and they cited the disclaimer in the deed of sale when the landfill went to the Niagara Falls school board. Friedman told the Hooker representatives that, since the company had placed chemical materials in Love Canal and in the 102nd Street dump, in his opinion the company *was* responsible and should make plans to remedy the situation. Soon afterward, representatives from Hooker and from the city of Niagara Falls met with Friedman again, telling him that they would solve the problem jointly. The DEC official responded that, if the city wished to share the responsibility with the company, it was their prerogative and was a matter between the city and the company (Friedman 12/1/78).

During the spring of 1977, the regional office of DEC felt hampered in trying to assess the extent of the problem, for they had neither the sophisticated equipment nor the expert personnel necesary to analyze chlorinated hydrocarbons thoroughly (Friedman 12/1/78). The DOH laboratory "told Buffalo officials it was too swamped with Mirex and PCB analysis from the Hudson River to comply with their repeated requests" (Pollak 5/1/77). Although people complained about the fumes in their basements, there was no plan by state or county officials to collect and analyze air samples. No one had yet tried to examine the possible links between fumes, materials in sump pumps, and the health of residents (Friedman 12/1/78). In May 1977, however, the county health officer who had previously suggested that residents should prevent leachates from draining into the river announced that the health threats from the seepage were slight (Pollak 5/3/77).

In April 1977, the city had hired the Calspan Corporation to develop a program for reducing the groundwater pollution at Love Canal. One month later, the city knew the problem was extensive. The Calspan survey showed that 21 of 188 homes adjacent to the canal had varying degrees of chemical residue in their sump pumps and strong odors, with 75 percent of the affected homes clustered at the southern section of the Love Canal site (Pollak 5/1/77). The report prepared for the city by Calspan stated that storm sewers to the west of the canal contained the dangerous substance PCB (polychlorinated biphenyl). Furthermore, investigations of the large fields next to the school, concealing the old Love Canal:

> . . . revealed that areas of the Love Canal containing drummed residues have corroded drums at depths of 3 feet or less with many being exposed at the surface. Organic material from the drums is sometimes visible on the surface or is encountered within a few feet of the surface . . . malodorous fumes from the leaked materials are perceptible at all times. (Leonard, Wertham, and Ziegler 8/77, p. 20).

According to the New York State Department of Health, that report was not made available to them until late March 1978 (NYDOH, 1979, p. 4). The lack of communication of important information between agencies and levels of government was to characterize their relations as events continued to develop.

Local news coverage began to pick up again in the summer of 1977. Early in August a *Niagara Gazette* reporter wrote a front-page story summarizing and updating the Love Canal problem and detailing the cleanup plan submitted by the Calspan firm to the city. (Brown 8/10/77). A few days later, that newspaper's editor urged the immediate start of the $425,000 project, proposed by Calspan, to seal off home sump pumps, install a tile drainage system, and partially cover the leaking canal site (*Niagara Gazette* 8/15/77). Within a week of the editorial, however, the city officials decided not to assume the responsibility for the cleanup work recommended by their consultants (McMahon 8/9/78). If anything, the flurry of publicity may have done nothing more than lower real estate values in the neighborhood, according to some residents, thus adding to their anxiety.

In this context of inaction in the face of expert opinion that there were dangerous chemicals in people's dwellings, a Niagara Falls employee contacted U.S. Congressman John LaFalce, urging him to visit the area. In early September 1977, Congressman LaFalce toured the area. The things he saw and his conversations with numerous homeowners disturbed him greatly (LaFalce 12/24/78). He wanted the city to act, and he advised acting city manager, Harvey Albond to write to the Hooker Corporation asking them for every detail about dumping practices—when and where and what had been dumped. The city manager responded that there might be legal problems, since Hooker had not owned the land for many years and parts of the area were privately owned. In 1955, the school board had sold 6.58 acres of the northern end of the property to the city of Niagara Falls. In 1962, the board sold 5.98 acres at the southern end to a New York City resident, who, in turn, sold it to a man who lived in Pennsylvania. The congressman then asked a water treatment plant employee to send a report to him within a few days. When no report appeared on the due date, the congressman learned that the city manager had told the employee not to talk with LaFalce and to clear anything said about Love Canal through him first (LaFalce 12/24/78). The employee had spoken freely with the press previously, but this was not to be the last time city employees were restricted by their employer in regard to Love Canal.

LaFalce was accused by a local city councilman of "grandstanding" when he criticized the city for not taking adequate measures to correct the Love Canal situation (*Niagara Gazette* 9/12/77), but he was not to be stopped so easily, then or later. He wrote to the Hooker company and to the federal Environmental Protection Agency (EPA), asking them to send someone to visit the area. In late September, members of EPA's regional office visited the site with LaFalce, who was hoping for some financial aid from EPA so that remedies for the problems could be initiated. During the visit, the EPA's acting regional administrator was impressed with the seriousness of the situation (LaFalce 12/24/78). The regional EPA director soon recommended, in part, that "serious thought should be given to the purchase of some or all of the homes affected, or at least of those willing to sell" (Moriarty 10/18/77). The possibility or necessity of moving people out, which became a major issue later, was evident to observers from that date if not before.

At about that time, in September 1977, Friedman, the regional DEC director, made a referral to the DEC attorney for legal action against the Hooker Corporation (MacClennan and Shribman 8/15/78). The case was forwarded to the compliance counsel in Albany, where a decision was made to incorporate the Love Canal issue into one large legal action with issues of other Hooker disposal sites. The regional director was dismayed—worried that the Love Canal matter might be lost in complicated, extended litigation, with no relief for the affected people. By December 1978, he commented that "nothing has been done about Hooker. Every time I ask, I get answers that don't really answer the questions" (Friedman 12/1/78; see Brown 12/10/78). It was not until April 1980 that Robert Abrams, attorney general for New York State, filed a $635 million lawsuit against the Occidental Petroleum Corporation and Hooker Chemical Corporation. Among other things, the defendants were charged with negligence in preventing the migration of wastes and in failing to warn the public of the Love Canal hazards (MacDonald 4/28/80).

In October 1977, after a request from the regional office of the DEC, the EPA engaged a consultant to sample air in the basements of Love Canal homes (McMahon 8/9/78). That there was some problem with the air was very clear to some area residents and now to state and federal agencies, if not to local officials.

Although exploratory and preparatory actions had begun in 1976, little change was visible by early 1978. Pointing out that six months had passed since the city's promise to act within six weeks, Mike Brown ended a *Niagara Gazette* column in February 1978 with the scathing comment that state and federal agencies should recognize that they were in the midst of a true environmental crisis and "should get cracking on this problem . . . or take words like 'protection' and 'conservation' out of their department titles" (Brown 2/5/78).

During the early spring of 1978, the EPA continued to sample air in home basements, while the regional DEC personnel sampled basement sump pumps and the storm sewers next to Love Canal, whose waters ran untreated into the river. At that time, the EPA hired the Fred C. Hart Company to study the groundwaters in the Love Canal neighborhood, and their report was submitted that summer, confirming the need to correct the hazardous situation quickly (Hart 7/28/78).

It was also in the early spring of 1978 that the state health department's director of laboratory sciences, Dr. David Axelrod, learned that the soil samples his laboratory was analyzing for DEC came from a residential area. Alarmed, he insisted to Health Commissioner Robert Whalen (whom he would succeed within a few months) that serious public-health problems might well exist at Love Canal (Brown 10/22/78). On April 13, 1978, Dr. Whalen and the commissioner of environmental conservation, Peter Berle, flew to Niagara Falls to examine the site. Within two weeks, Dr. Whalen stated that the conditions at Love Canal represented a serious threat to the health and welfare of residents, and he directed the Niagara County health commissioner, Dr. Clifford, to undertake a series of corrective actions (McMahon 8/9/78; Whalen 4/25/78).

Now under pressure, the Niagara County Health Department acted. Dr. Clifford told the chief officials of the city of Niagara Falls to clean up the most visible signs of chemicals—the drums emerging at the surface of the old canal site and the standing puddles of chemical-laden water. The city officials passed the order on. Although Dr. Clifford reported to his superior, Dr. Whalen, that the exposed drums would be covered and exposed chemicals removed or covered (Clifford 5/10/78), an official recalled that the task was not as simple as putting "some dirt out there." The city engineers told him that, if he simply put dirt on the puddles, the liquids would run out into the streets. Residents recall, however that dirt was put out on the canal from time to time in an unsuccessful attempt to cover some of the open places, and also that a few "fifteen dollar fans" were placed in some homes to help with the ventilation of the basements (Brown, 1979, p. 21).

Visits to the neighborhood by high-ranking government officials, who even talked with residents and saw some homes; the testing of soil, sludge, and air in and near the homes; the detailed reports in the newspapers—all alerted the Love Canal residents, alarming some but making others hopeful that help was at hand to solve problems that they had already discerned. Although the agencies were finally paying attention to the problem, their actions seemed intolerably slow to people living every day in sight and smelling distance of chemical-filled holes in the ground, worrying that, when the weather grew warmer, things would be worse, especially for those families whose children would be out of school for the summer.

Corrective measures were undertaken at a creeping pace. One of the actions Commissioner Whalen ordered the county commissioner to undertake in April was installation of fencing to restrict access to the canal site (*Niagara Gazette* 4/28/78). It was not until June or early July, after yet another directive to the county (Whalen 6/20/78), that a flimsy snow fence was placed around part of the canal area to indicate that it was dangerous (McNeil 8/2/78). "No Trespassing" signs were put up as well. However, children were playing on the school grounds all summer, under a supervised playground program, until mid-August 1978 (*Niagara Gazette* 8/5/78*a*). Finally, the playground topping the old canal was declared out of bounds, but it was not out of reach for children, who ducked around the fence and ran across the big field to friends' houses or played games as usual in the big open space.

More and more, as the residents discussed some of their fears for their health and financial security, they realized their problems were shared ones and not of their own doing. They also talked freely to sympathetic reporters. A few neighbors whose properties abutting the canal site showed distressing signs of the chemical presence—an inground swimming pool that "popped up" two feet, holes with chemicals, dying shrubbery—began to collect names on a petition to city hall. They requested relief from taxes on homes that were dropping sharply in financial value, through no fault of their own (Brown 5/21/78; Swan 1979). As the citizens organized, and as news coverage increased, pressure was thus exerted on elected officials during that spring and summer to do something about the problem.

Government Officials and Residents: The Beginnings of Mistrust

In the second week of May 1978, on the day that the EPA concluded that the toxic vapors in people's basements suggested a serious health threat (Brown 5/15/78; NYDOH, 9/78), state officials met with Love Canal residents to ask for their cooperation in studies to assess health and environmental problems. A second meeting a few days later was a fiasco. Dr. Clifford, the County Health Commissioner, said his department had failed to carry out all the directives from the State Health Department. He explained that his examination of 99th Street School attendance records assured him there were minimal health hazards, and he said property values were the homeowners', and not government's problem. Dr. Stephen Kim, a toxicologist from the New York State Division of Laboratories and Research, heatedly disagreed with Dr. Clifford, on the stage (Kim 5/22/78; Friedman 5/23/78; Violanti, 5/19/78). As the meeting progressed, residents received little precise information, except that one family was informed it would encounter health risks if anyone stayed in the basement for more than 2.4 minutes (Brown 5/20/78). A few days later, a state biophysicist was quoted as saying that if he lived at Love Canal and could afford to move, he would do just that (Brown, 5/25/78).

During the spring and summer of 1978, more people became aware of the problem of Love Canal. Residents attended meetings; listened to the statements by officials, who either denied problems existed or pointed with alarm and then offered no help or advice; listened to neighbors' accounts of the meetings; and read the comments about their homes and neighborhood. They became terrified that their homes—especially their basements, where many had innocently built bedrooms and family rooms—were dangerous places.

What frightened people even more, however, was the possibility, first, that no one in authority knew what was really wrong, and, second, that even if they *did* know exactly what was wrong, maybe no one would help anyway. The residents already knew that the problem was too big for individuals to solve. Since they could not sell their homes, for there were no buyers waiting to purchase them, the residents knew that, if they moved, they would still be faced with monthly mortgage payments and maintaining their homes, in addition to renting or purchasing other places to live. Their choices were to face financial ruin in an uncertain situation, or to hang on and hope either that the conditions would be completely corrected or that somehow they would be given some financial assistance by some level of government.

The residents who were becoming concerned were distrustful and disdainful of city and county officials but hopeful that state officials would take heed of their plight and help them. They had not yet had much contact with federal agency officials. Based on their experiences with Congressman LaFalce, and knowing the vast resources of the federal government, they were certain that if only both the federal government and the state would become more involved, then surely help would be forthcoming. That involvement would not happen for some time, however.

Newspapers continued to alert the people throughout May 1978, and by early June the people were talking to each other more and more, thinking about "getting together somehow," and they were far more vocal at public meetings with officials. When a meeting was held in June to inform the residents about engineering plans to correct the leachate drainage at the southern, more obviously troublesome end of the canal area, the residents had many specific questions, but the answers seemed vague and unsatisfactory to an audience that included blue-collar workers experienced in construction work of all sorts, and they were skeptical of assurances from engineers that the plans and calculations were foolproof (see USHR 3/21/79, Gibbs testimony, pp. 87-92).

The mistrust between citizen and government, between lay person and expert, took hold when people began to doubt not only the motives but the wisdom of the bureaucrats and their scientists and consultants.

As they thought about the various illnesses in their own families and became more informed about their neighbors' illnesses, as they read the reports in the newspapers about illnesses related to the chemicals that they were learning were buried nearby and were possibly present in their yards, basements, and air, they became increasingly concerned. Reassurances from official sources made them feel more and more that only people living in the Love Canal area could really understand what was happening.

Good intentions on the part of concerned officials were not to suffice as substitutes for a solid understanding of the area's geography and history and of the experience of the problems from the residents' viewpoints. Once the first epidemiological studies were launched, the researchers found themselves swept up in social events that were unpredicted and costly.

In late June 1978, the medical investigators drove a van down the streets directly adjacent to the landfill. Armed with a list of residents, the investigators took blood samples and left complicated health questionnaires at the houses whose backyards were next to the canal surface. For the first few days, the health-information collection procedures were orderly. Soon, however, residents across the street from those being tested, complained that they wanted tests, too, and then residents on adjoining streets also grew concerned. The state researchers did something that was to occur more than once. They reacted impulsively to loudly voiced complaints, immediately announcing that blood specimens would be taken at the 99th Street School for all of the hundreds of residents, and that people could pick up questionnaires at that time as well.

Though the reason for this decision may have been a desire to serve and to reduce people's anxieties, the disorganized response was an indication to people of a lack of a real plan, which added markedly to people's caution and lack of respect.

By this time, the public meetings were beginning to serve a function that had not been intended by the state officials who convened them, under the rubric of "citizen participation," in order to convey information to the residents and to gain their cooperation. The cooperative participation did begin, but it was between citizens who were growing aware of one another and their common concerns. Even at this early stage, though interactions with the officials were mostly cooperative, they also began to include strong elements of confrontation. The public sessions also provided information to officials who were willing to listen; they provided emotional catharsis for the increasingly impatient residents, and they attracted the excited attention of reporters.

By August, the state health department had collected more than 2,800 blood specimens and eventually collected more than 4,000. This massive collection strained the resources of state laboratories, which further confounded residents. "Why did they do too many at once, then?" was the residents' exasperated response.

The stirrings of suspicion that the experts were confused increased when the state began to issue the results of the air-sampling survey started during the last week in June. Starting early in July, many people were apprised of detectable levels of industrial chemicals in the air of their homes. They received mimeographed sheets listing values of air readings taken in their basements and other rooms. Numbers were listed next to the chemical names: benzene, toluene, benzoic acid, lindane, trichloroethylene, dibromoethane, benzaldehydes, methylene chloride, carbon tetrachloride, and chloroform. The numbers and chemical names were listed with the street address but with no explanation. When worried residents asked for interpretations of the information, the state officials acknowledged that they did not know. In a meeting later in the fall of 1978, when asked why people had been given raw figures, a state official replied that the state wanted to share everything with the people!

Whether this was said sincerely or as a mocking concession to the residents' repeated demands that they know about and share in decisions about their lives, the reply and the entire interchange typified the relationship that began to develop early in the summer of 1978, when the state agencies slowly assumed responsibility for the Love Canal problem. The more that officials met with residents, the more negative feelings and relationships developed. When professionals presented raw data, it confused people. When they tried to interpret the data in down-to-earth terms, describing risks as some number of deaths in excess of the usual number expected, people interpreted that to imply *their* deaths and their children's deaths. When they tried to calm people by saying that, despite all the serious possibilities, there was no evidence of serious health effects, the officials were seen as covering up, since no health studies had been done. Authorities trying to coordinate multiple governmental and private agencies were seen as wasting time in meetings. What the officials thought of as privileged advisory conferences were viewed as conclaves that excluded affected citizens. What officals saw as preliminary studies conducted to assess the situation were viewed by residents as wasting resources on repetitious research projects rather than doing something helpful. When they took action quickly or tried to do everything at once, for everyone, they overloaded facilities, made errors, and were faulted for bungling. When someone would offer a genuine, off-the-cuff expression of sympathy—"I sure wouldn't live here if I could avoid it"—it seemed to help no more than did the "stonewalling" of the old-line bureaucrats.

Always, woven into every contact and influencing every decision, was the question of who was going to take responsibility, who was going to pay actual dollars, and who would commit resources of personnel. This unsolved question made decisive action impossible for the first few months. In December 1977, for example, Frank Rovers, a new consultant, had been brought in to advise about covering the canal and draining excess leachates from it. He addressed a Love Canal study group whose members represented the city of Niagara Falls, the school board, the county health department, Hooker Chemical, and legal counsels for the city and for Hooker. An air of harmony reigned when Dr. Clifford began the proceedings by saying he preferred to take care of the problem by cooperative rather than coercive means. Rovers was promised $5,000 so that his firm, Conestoga-Rovers, could devise an engineering plan.

Once the plan was complete, however, the study group began to fall apart. Late in June 1978, the school board announced that it had deeded the land to the city in 1960, that it was not responsible for the health hazards and property losses, and that it would be short of cash as a result of a recent court decision limiting its ability to tax over state constitutional limits. In July, the city announced its withdrawal on the advice of the city's bond counsel, who was concerned about private ownership of part of the land (Seal 7/28/78). The Hooker Corporation had helped finance the preliminary engineering study and had provided maps of the dump site to state officials. However, the company stated that it was just acting as a "good citizen" and was not accepting any responsibility for the planned major construction project (Brown 6/20/78*b*). There now arose a possibility that the U.S. Army had disposed of materials in the old Love Canal; city authorities were reported hopeful that this might lead to federal financing of remedial measures (Brown 6/23/78). On the request of Congressman LaFalce and Dr. Axelrod, the Defense Department conducted a brief investigation of reports by residents who claimed they had witnessed the dumping of wastes from Army trucks decades before. The Pentagon soon issued a denial of any such actions (Hildebrand 8/14/78) and, despite Congressman LaFalce's angry protests to Secretary of Defense Harold Brown (LaFalce 6/26/78), there was no change in that stance.

In this context—all other governmental entities and the Hooker Corporation ignoring the situation or denying responsibility for the consequences of the long-ago burial of chemical wastes—the state of New York seemed to emerge as the one body with the authority, the mandate, and the resources even to try to do something about the problem.

By mid-July, 1978, the residents had had their fill of news announcements, official inquiries, meetings, work days missed, and long, feverish discussions with spouses and neighbors about Love Canal. At best, those who were keenly aware of the problem were nervous and angry about their lost summer; at worst, they feared that something ominous, huge, and destructive might be in store, disrupting not just one summer but many seasons to come. Now they waited impatiently, wanting and yet dreading to hear what the commissioner of health would announce on August 2. There was to be a meeting to decide, the rumors went, whether people would be evacuated from homes near the southern portion of Love Canal and what, in general, was going to be done about the entire problem. Everything in this long, hot frightening summer seemed suspended while people waited for the decisions to be announced by the health commissioner at a meeting in Albany—300 miles from the affected people and their polluted neighborhood.

No one knew then that the decisions to come would catapult the situation into the national and international consciousness, and that Love Canal would stay there for years to come.

Notes

1. I have received no reply to my queries to Donald C. Baeder, president of the Hooker Corporation, about these points (Levine 7/22/80, 9/9/80).

2. According to the Niagara Falls City Engineer's Department, the streets were completed as follows: Colvin Boulevard, November 1958; Read Avenue, September 1960; Wheatfield Avenue, May 1956; Frontier Avenue, October 1970; 97th Street, December 1959; 99th Street, May 1956.

3 "You're Murdering Us"

By August 2, 1978, Robert P. Whalen, commissioner of health for New York State, must have felt fully prepared to make his official pronouncement about the Love Canal situation. He had visited Love Canal once, in April. With his staff, he had examined the results of preliminary epidemiological and environmental studies conducted not only by his own department but also by the U.S. Environmental Protection Agency (EPA), their consultants, and two private engineering firms, Calspan and Conestoga-Rovers. In addition, the commissioner and high-ranking members of the department of health had met on three occasions with other agency members and their consultants. On June 15 and July 14, they had met with representatives from the state department of conservation, the EPA, and the offices of Congressman LaFalce and State Legislators Murphy and Daly, as well as with the Niagara County Health Commissioner. There also had been a six-hour meeting at LaGuardia Airport with expert consultants from all over the nation: physicians from the National Cancer Institute, the Centers for Disease Control, the National Institute of Child Health and Human Development, and other experts in toxicology and industrial hygiene from several state universities and large corporations (Gunby, 1978; Whalen, 8/1/78).

Although the residents had been *informed* at some public meetings of the intentions of the department of health, some had been interviewed informally, and some had completed health questionnaires, they were not *consulted* as people who might have important contributions to make about the history of the area or their own experiences or might have opinions about what should be done to and for them. Nor would they be consulted in the months to come, as the experts assumed that all the relevant answers could be discovered readily and that they knew all the relevant questions to ask. Not even the names of the eminent consultants were revealed at this time—a policy repeated when decisions crucial to Love Canal residents were made during the following months. What the residents viewed as secrecy and their exclusion from vital decision making was to result finally in their suspicion that there was a cover-up.

But on August 2, 1978, the commissioner was ready to share his decisions, in the form of a health department order, with "all interested people." The auditorium, however, was far too large for the audience of some fifty people.[1] A disquieting rumor had gone around the Albany offices that 160 Love Canal residents were going to attend the meeting. In fact, only a handful were there, not too surprising a showing, since driving the 600 miles round trip between Niagara Falls and Albany meant a lost day or two of pay, and flying would take a large proportion of the average week's salary for a Love Canal resident.

Health Commissioner Whalen proceeded to read his order. He announced:

> [The] Love Canal Chemical Waste Landfill constitutes a public nuisance and an extremely serious threat and danger to the health, safety and welfare of those using it, living near it, or exposed to the conditions emanating from it, consisting, among other things, of chemical wastes lying exposed on the surface in numerous places and pervasive, pernicious and obnoxious chemical vapors and fumes affecting both the ambient air and the homes of certain residents living near such sites.

There were ten selected compounds, he explained—seven of them carcinogenic to animals, one a human carcinogen—all found in some of the homes tested, in the form of vapors. A study of residents living near Love Canal, "utilizing spontaneous abortions and congenital defects as indicators of potential toxicity," showed that there was "a slight increase in risk for spontaneous abortion among all residents of the Canal," with the highest risk among those living at the southern end, on 99th Street. He concluded: "A review of all of the available evidence . . . has convinced me of the existence of a great and imminent peril to the health of the general public residing at or near the said site as a result of exposure to toxic substances emanating from such site" (Whalen 8/2/78a).

Stating that, pursuant to his authority under the Public Health Law, he was declaring "the existence of an emergency," he ordered and directed that nine measures be undertaken to preserve and protect the public health. The two measures that were most important in their long-range and social consequences were that the city of Niagara Falls and the Niagara County Board of Health were to implement the Conestoga-Rovers engineering plan to stop the migration of toxic substances and that the county health department and the city, in cooperation with the state department of health, were to undertake studies to determine the extent of chronic diseases that afflicted residents living adjacent to the Love Canal landfill. They were also to undertake studies to determine how far the chemical leachates had migrated from the canal and whether the groundwater aquifers had been contaminated. In connection with the Conestoga-Rovers engineering project, he ordered that a plan be developed for the safety of workers and "to minimize hazardous exposure to residents" during the course of the construction work. Dr. Whalen also ordered the 99th Street School to be closed temporarily.

He also made seven recommendations. Two of these were addressed to all Love Canal area residents, telling them to avoid going into their basements and to avoid eating anything from their gardens. These two recommendations alarmed people considerably, for many, seeking to reduce food costs, had depended for years on the home-canned product of their own gardens. All residents used their basement for laundry, workshops, and other purposes; many had built bedrooms and playrooms which were in daily use. They wondered, too, how contamination could be confined to one area and not another. If the basements and gardens were unsafe, they reasoned, how could the remainder of the house and property be considered safe for their families, particularly since the health commissioner ordered that studies were still necessary to determine where the leachates had migrated and what sorts of damage they had done?

Two other recommendations were destined to become the center of a storm of controversy, both immediately and in the long run:

> (1) That pregnant women living at 97th and 99th Streets and Colvin Boulevard temporarily move from their homes as soon as possible; (2) that the approximately twenty families living on 97th and 99th Streets and Colvin Boulevard arrange to relocate temporarily any children under two years of age as soon as possible (Whalen 8/2/78a).

The wording of these recommendations was soon changed. The first was swiftly amended to include *families* of pregnant women (Whalen, 8/2/78b), but the change came after a day or so of excited confusion for the residents.[2] Regarding the scope of the recommendations, evidently Dr. Whalen thought the affected group would number less than twenty-five families (Whalen 8/1/78). This number was to change drastically over the following days.

A local official in the city waterworks, thoroughly familiar with the local sentiments, whereas the Albany-based officials were not, remembers: "We heard he was going to recommend that pregnant women and children under two move . . . no money, just move. I couldn't believe it. I thought, Oh, God, what is going to happen?"

From the viewpoint of the hard-pressed commissioner, however, and those who had worked so diligently to this point, what did happen next must have been a shock—disappointing and puzzling. As a physician in a respected professional position, Dr. Whalen must have expected people to trust in the wisdom of his carefully considered decisions, prepared after consultation with all varieties of experts (except the people undergoing the experience). When the commissioner finished intoning his formal order to the hushed audience, however, one of the few residents who attended the meeting shouted "You're murdering us!" Although that shout was not recorded by the television and newspaper reporters, who had already left to file their stories, her comments were not to remain hidden for long—nor was she.

The young woman who shouted was Lois Gibbs, who would mold and lead a strong citizen's group and who would become nationally known for her battle with government over the next few years. At that time, however, she was still very uncertain how to proceed. Early in June 1978, she had read a news article in her local paper that "really upset" her. It stated that the chemical substances present in Love Canal were known to be associated with respiratory and neurophysiological disorders. She had been only mildly interested in the situation before, because she lived "a safe two blocks away" from the canal site, but now her attention was riveted. Since her six-year-old son had started attending the 99th Street School, he had developed severe asthmatic symptoms and convulsions. Upon reading the newspaper article, she made the mental connection between the events.

Lois Gibbs had married shortly after completing high school; her main concerns were her family and home; baby care, sewing, and cleaning filled her days. Social clubs, church activities, PTA, and political work were of no interest to her. She now found it hard even to think of how she might try to protect her child from vague menaces—from chemicals with tongue-twisting names—but she was no longer ignoring the situation; she wanted to learn all she could and, most of all, she itched to *do* something for her little boy.

She and her husband, a chemical production worker, related their fears to her sister Kathy's husband, Wayne Hadley. Hadley, a biologist and a university faculty member at SUNY/Buffalo, was keenly interested in environmental issues. When he was told about the news reports, he became alarmed. Not only was Hadley professionally concerned about environmental hazards, he was interested in the welfare of his wife's family; moreover, his six-year-old son spent many days with Lois, who took care of the child during Wayne and Kathy Hadley's working day.

Hadley suggested that Lois request that her son be transferred to another school. The school principal told her to submit a physician's letter along with her request. When she complied with not one, but two letters, the board of education refused her request on the grounds that there was no chemical hazard at the school; therefore, there was no reason to transfer a child whose physicians stated that he must not be in a school with chemical vapors present.

The official denial angered the young mother; she was not to be put off so easily. She decided to look for help from fellow parents. She telephoned the PTA president, who told Gibbs that she did not believe such a problem was within the PTA's purview. Then Lois called Karen Schroeder, a Love Canal resident from childhood, now married and the mother of a child with multiple birth defects. The Schroeders had suffered severe property damage, dating back to 1974, when their inground swimming pool suddenly rose about two feet out of the ground from the pressures of underground waters. When the pool was removed the next summer, the hole it left immediately filled with chemical-laden waters, the shrubbery on their property soon died, and their showplace of a yard became a wasteland. Her parents had been vainly trying to clean black sludge from their basement walls since 1959. Karen had complained to City Hall, had attended city council meetings to voice her complaints, and had helped to start and was active in the first citizens' group that formed about the Love Canal problem. That group was composed of some families living next to the canal who were seeking tax abatement and considering withholding mortgage payments. One important function Karen Schroeder had served was to work very cooperatively with Mike Brown (and, later, with other news reporters), relating her family's experiences to them and helping the reporters meet with other families who felt affected by Love Canal's contents (Brown 1979).

Karen told Lois that her group was organizing and was getting a lawyer (Brown 5/21/78), and that she would call Lois back in about a week. That group's concerns seemed somewhat different from her own, Lois decided. Now she did not know exactly what to do, but she wanted to do something. Her brother-in-law suggested that, if she really wanted action, she should "Get ten mothers down there picketing for ten days and you'll close the damn school down!"

The thought of doing something so bold was frightening to Lois. She continued to read the papers to try to understand what was happening, she visited the newspaper library, and she even tried to read whatever she could to learn something about toxic chemicals, "but there weren't many things at my level." She then decided to try to interest some people in forming a committee—"to get some sort of a cleanup of the school area, and in general to form a parent's action committee for whatever might need to be done to make the school safe for the children."

On the first day of her new endeavor she forced herself to walk to a house at the corner of the street where the school was located, and knocked on the door. To her great relief, no one answered. "I ran home, thinking, 'well, I tried!'" She stayed home for a full day, but once the initial feelings of relief had subsided, she realized that she must keep trying, even though she was "scared stiff at the thought." The first people she spoke to lived on 99th Street. They were not very interested in the problem. Some reasoned that, since they lived next to the canal anyway, it probably made little difference to their children's health if they were in the school for five or six hours a day. A few people on that street were somewhat more responsive and "no one really put me down." She then went to her own street, two blocks away, and found the people there more interested in what she had to say. Since they believed they lived fairly far from the chemical-landfill site, they seemed to think that the children's exposure for several hours a day could be a health risk.

Lois continued to knock on doors, slowly making her way up and down the street during the hot summer days. It took her four or five weeks to work through two blocks. She found people so eager to tell her about themselves and about their feelings, opinions, and concerns about what was happening that she was spending about forty-five minutes at each house. She was thus absorbing a good deal of information about problems from the people. She continued to read everything she could about pollution and about Love Canal in the newspapers, and she questioned anyone who had any information at all. In addition, she was consulting regularly with her brother-in-law. Most of the time, however, she was knocking on doors and talking with people.

She was learning a great deal, not only about people's problems but about her ability to respond to people and to put her emotions to work in a controlled fashion. She learned some things on her own and some, for a time, under the direct tutelage of her brother-in-law. When she attended a meeting held by officials in June 1978 to impart information about general matters and about the construction plan, she asked: "Is the school safe? Do you think my children should attend?" When there was no satisfactory answer, she said, "Get this school down, it's contaminated! What if the barrels come up out of the ground?" At that time, she was not quoted in the papers. On this first occasion of asking questions publicly, she had not planned her questions ahead of time, but rather spoke impulsively, emotionally. Her brother-in-law pointed out that she had not fully used the opportunity inherent in asking questions in a meeting—to reach a public larger than those present. He advised her to have a good set of questions ready and to state them clearly and forcefully within the first fifteen minutes at any meeting, "because the media stays fifteen minutes and they get out to file reports." (Later, when television news cameras were trained on her regularly, she learned to make her points within fifteen seconds.)

Wayne Hadley provided far more help than this, however. He had been actively involved in environmental skirmishes in the past, and he was concerned about the role of industrial polluters of freshwater lakes and rivers. Once Lois became active, Hadley devoted every moment that he could spare that summer to the efforts of Lois and the people who began to work with her. He attended meetings with Lois, interpreted situations for her, described the government structure he was familiar with, explained some minimal principles of organizing people, criticized her ideas and her presentations of them, gave her reading materials, and put her in touch with other environmentalists, who provided some support in the days to come. In general, he tried to prepare her and others for what he sensed was going to be a long, and very tough battle.

On Hadley's advice, when she had the names of about thirty interested residents, Lois and two other women visited Richard Lippes, a young

Buffalo attorney who was well known locally for his interest in environmental issues. In fact, Lippes had just been named president of the local chapter of the Sierra Club. When the women asked him if he would be interested in bringing some sort of suit for all the people, he replied that he would look into the matter and get back to them.

At the end of the fourth or fifth week, Lois was "really beginning to feel it." She was physically exhausted, neglecting her house, leaving the children with her husband and mother, and uncomfortably aware that her husband missed her and their normal patterns of living. Most of all, she was feeling weighted down by her accumulating understanding of the enormity of the problem. She was getting "not discouraged but frustrated," as she realized the scope of the problems people had been living with, each family unaware that they were surrounded by other people with serious-sounding physical problems. One day, just as she felt she was getting too tired for the huge task, she was working on a street adjacent to the canal site. There she spoke with Debbie Cerrillo, and when Lois told Debbie what she was doing, Debbie became keenly interested, for her home was in one of the worst areas, and large, foul-smelling puddles were ever present in her backyard. Lois asked her to help. Debbie thought about it, talked it over with her husband, and told Lois she would be very interested in working with her. Now there were two people to go door-to-door, and things became, if not easier, at least shared. Lois now had an ally, who remained a loyal lieutenant in the battles to come, and Hadley now had two eager pupils.

By the time Lois and Debbie teamed up, thirty-nine people had indicated an interest in being a part of a local residents group. Debbie and Lois decided on a three-part purpose for the group they would ask people to join: to press for some restitution for property losses, to insist on a cleanup of the chemicals, and to work for the immediate closing of the school. They asked people if they were willing to have Lois and Debbie represent them in talking with governmental authorities. As community members, they understood some subtle points outsiders might not have been sensitive to. They did not ask anyone to sign anything, for example, because they knew that people sometimes "feel leery about signing things." They wrote people's names on a list, however, and in exchange gave them "tickets" they had made from pieces of paper torn from a spiral notebook. On the tickets were their names and phone numbers and the words *Love Canal Homeowners Committee.*

The word *Homeowners* was chosen rather than *residents* because Lois was advised that, if they wanted to have tax or mortgage protests, they would have to be a group of property owners, and that property owners would be listened to more readily than nonowners. Later, the name added to the problems of trying to urge the renters living to the west of the canal area to join the organization. (The word *Homeowners* seemed to emphasize the differences in core interests, which kept the owners and renters from ever forming more than an uneasy coalition.)

Lois was relieved to have Debbie working with her. She had never been totally alone with her work—her husband, brother-in-law, mother, and sister all had worked hard, too—but now she not only had someone to share the door-to-door work with, she also had a companion in addition to her husband to go with her to Albany, where they would represent some part of the Love Canal community at the August 2 meeting, where the commissioner of health was supposed to meet with interested parties to work out decisions about the Love Canal situation.

Lois and Debbie decided to undertake the trip to Albany after they were barred from a meeting in Niagara Falls between city and county officials because they had no official standing other than as affected citizens (Stutz 7/1/78). By the time they and Lois's husband, Harry, left for the state capital, they had the names of more than one hundred people willing to be represented by their new committee, which was soon to become an association. Although they were rather innocent at that time about public officials and about how to conduct themselves in a sophisticated world, they learned very quickly; they bolstered each other through this first experience, and they were to prove both tough and tenacious in their quest for what they saw as just treatment for the Love Canal people. When they finally arrived in Albany, they slept briefly and were present when Dr. Whalen opened the meeting. Much to the surprise of the residents who had come—Lois, Debbie, Harry, and a contingent of four from the tax-and-mortgage-action group—Dr. Whalen proceeded to read a formal order. The residents had assumed that they had traveled all the way to Albany to *participate* in the decisions to be made about Love Canal—about their homes and their health—but they quickly realized, as Lois Gibbs said later:

> All the decisions had been made. In fact, the final decisions had been made no later than nine o'clock of the previous evening. We were very perturbed . . . they made a recommendation . . . they had no financial backing to move these people.

When Lois realized that they had made a tiresome, expensive trip to hear an order read, an order that seemed not to address the problem completely, she shouted, "You're murdering us!"

Once Lois had spoken so strongly, Debbie stood up, and asked, "What about my two-and-a-half-year-old; she's out of luck, right?" At this meeting, the two young women began to flex some muscle. The meeting, they learned later, was supposed to last for a half-hour, but they persisted in asking questions about every point, keeping everyone there for more than an hour—except the commissioner, who left immediately after reading his order and hearing the first "citizen reaction."

One final remark, which still made Lois angry when she thought of it months later, was the health commissioner's statement that the health department had not known anything about the situation until about ninety days previously and that the department could not do anything about problems they were unaware of. The ignore-deny pattern, couched in official obfuscation, had now extended to the state level.

Finally, the meeting over, Lois, Harry, and Debbie drove the 300 miles back to Niagara Falls. They were in hurry to return to their children, and they were angry and disappointed at the commissioner's recommendations and the manner in which the recommendations had been devised. At a rest stop, Debbie phoned her mother and learned that the decision to close the 99th Street School had already been announced in Niagara Falls. That victory was fading in the young women's minds, however, before the implications of the health commissioner's order to recommend a move of pregnant women and children under age two.

Those implications—what would happen *after* decisions were made and announced, what the affected residents' reactions might be—had not been anticipated by the health commissioner or his many advisors. People's reactions to decisions were simply not viewed as appropriate parts of the complex of problems to be addressed. "We deal with physical facts, not with social and political matters," one health official mentioned later. Not only that agency repeatedly ignored the social facts of people's reactions, however; so did virtually every governmental body that became involved over the next two years. Lois and Debbie were keenly aware of the anxiety and anger that had been building in their neighborhood, and they wondered, somewhat apprehensively, when they would be able to get together with the people they had been talking to at their front doors. They soon learned they would not have to wait long to talk with their neighbors.

The news from Albany had preceded them; it had been announced on the radio, on television, and in the local newspapers while they drove across New York State. Since there were no officials available in Niagara Falls to meet with people, the residents, in their fear, had run from their homes to seek information, comfort, and directions for action from each other. It was pandemonium. The residents had been alerted to come to 99th Street for a meeting organized by the small tax-and-mortgage-protest group. They were planning to collect mortgages and burn them publicly to show how valueless their homes were now. The man who organized the meeting, an amateur musician, had put his home microphone outdoors to use as a public-address system. When Lois, Harry, and Debbie arrived, the people began to murmur their names. The crowd quieted down a bit when Lois's brother-in-law took the microphone and introduced Lois as "someone you already know who has been doing a good job going from house to house."

Until now, Lois had thought of herself as a painfully shy person. The *thought* of addressing a large crowd, particularly an angry one, "would have had me scared stiff." The exigencies of the moment demanded it, however; she had no choice, and she simply had to assume the role that she was suddenly playing.

So she started to address the crowd, almost immediately scolding them for telling *her* about crib deaths and malformations and not putting that information on the health department questionnaires. She then proceeded to tell them what they already had read or heard about Dr. Whalen's official order. She urged people to start filling out health department questionnaires completely, and closed by telling them to call her if they needed help.

Lois had traveled the route from uninvolved individual to new leader of a newly aware community with which she was very familiar. She had personally experienced a problem, had been unable to get satisfaction through established means, had shared her experiences with others, and had discovered that her family's problem was a common one, was large, and was the fault not of individuals but of external forces. She had already collected a great deal of information, and she had benefited greatly from the tutelage offered by Hadley in his almost-classic role of intellectual adviser to disgruntled working-class people. More important, she was rapidly learning that, consciously willing or not, she was going to be a leader, and she was learning what would be demanded of her in that new role. She also was learning that she could rise to the challenges. The phone calls all night long, that night and for hundreds of nights and days to come, not only told her what concerned the people—and that she could respond to those concerns—but also showed that they accepted her and turned to her as their leader.

All the next day, people gathered in small groups on the streets and in yards, talking about what was happening, trying to make plans. That evening there was a meeting of state, county, and local officials, attended by about 600 Love Canal residents. The meeting was chaired by Thomas Frey, director of state operations for Governor Carey. He announced that the state might help pay for the evacuation of the families with pregnant women or children under age two. The meeting was in chaos. About a quarter of the way through the meeting, a state health department official walked over to Lois and quietly said, "You've got to calm these people down. We can't get anything done." Lois went to the microphone and was introduced by Tom Frey. She said,

> Hi, everyone knows me, I'm Lois Gibbs, and they've asked me to calm everyone down. I suggest that you sit quietly and listen to the questions and listen to the answers and then boo the hell out of them.

She got a round of applause—but a member of the 99th Street tax-and-mortgage-action group shouted, "You don't represent me, no one does."

Lois had been accepted as the people's leader by the state official chairing the meeting—further legitimating her new position—but discovered instantly that there would not be total agreement about accepting her as leader, that there would be outspoken criticism and opposition from some community members, and that she would have to deal with it, often publicly and with little preparation. At this point, she simply smiled and walked off the stage.

The news reporters, photographers, and television cameramen, trailing their gear and bathing the group in hot lights, recorded what was happening and inevitably became part of the scene themselves, while helping to turn the event into a nationally known and shared experience. Remembering the occasions when her questions and shouts from the audience had gone unrecorded, Lois now followed her brother-in-law's instructions to make her important statement while the television cameras were still there. "Where's Carey?" she shouted, "I'd be here if I was governor!" and a few minutes later, "You're treating us like the Titanic! Women and children first!" Other people shouted too, "Where's the mayor?" and "Where's Hooker?"

The issue was now joined. The state health officials thought they were offering help in this dreadful situation in a way that was reasonable, concerned, and within the department's legitimate purview and capabilities. In the words of a young man pointing to his pregnant wife, however, "The damage is done! My wife is eight months pregnant. What are you going to do for my baby? It's too late for my child" (*Buffalo Evening News* 8/4/78). In the words of the crowd, too, on the brink of becoming a mob, they wanted out. They wanted to be out of the situation somehow: some wanted to be out of the area; others wanted their neighborhood cleaned and authoritatively declared to be the safe place to live that everyone had always assumed that it was. However it was to be done, they wanted to be rid of this unforeseen calamity, one way or the other. "We want out! We want out!" they chanted. They wanted what they felt was their just due—they wanted *out—immediately*!

Notes

1. The people who signed the registry on August 2, 1978, included nineteen from the New York Department of Health, six from the New York Department of Environmental Conservation, five from the U.S. Environmental Protection Agency, two from the New York Division of the Budget, one from the New York State Senate Health Committee, one from the city of Niagara Falls, one of Congressman LaFalce's aides, two from Fred Hart Associates, one from Conestoga-Rovers, three from Hooker, one

from the *Niagara Gazette,* one unknown (illegible), and seven Love Canal residents (NYDOH 1979, Sect. 28).

2. Recommendation (2) was altered to refer to "the approximately twenty families living on 97th and 99th Streets south of Read Avenue, with children under 2 years of age . . . " The changes in recommendations seem to be the chief discrepancy between the August 2, 1978, order included in NYDOH 9/78 and the order read on August 2, and included in a set of documents collected by the N.Y. Department of Health in the spring of 1979 in NYDOH 1979, Sect. 28, p. 10.

4 The Events of August

Love Canal could no longer be ignored. The state of New York had explicitly shouldered major responsibility for the welfare of Love Canal citizens when Health Commissioner Whalen issued the order directing and recommending a series of corrective actions for the area. Important decisions for implementing that order were worked out by state officials very shortly after the anguished meeting described in the previous chapter. The decisions were made swiftly—shaped in response to anticipated resources, unanticipated public reactions, and the rhythms and pressures of electoral politics. They were not made in response to prolonged assessments of either the physical or the social situation at Love Canal. Once decisions were made, the official perceptions of the Love Canal problem were shaped by the decisions themselves, as we shall see. The course of events of August 1978 set the stage for the ensuing months of struggle between residents and their government, as the residents organized, defining their goals in response to their perceptions of their problems and in response to the state's actions—finding their methods in part by hard trials and errors and learning from daily experiences. In these processes, as citizens and officials grew further apart, each viewed the other as creating part of the problems at Love Canal, and they became, to some extent, opponents—measuring but not trusting each other.

This chapter will describe the governor's solution to the problem of moving pregnant women and children from the Love Canal area and the initiation of the remedial construction project. Chapter 5 will examine the social consequences of the department of health's approach to the studies of "chronic diseases afflicting all residents." Although the various selected events are necessarily described here separately, they were happening at the same time and were affecting the Love Canal actors not in the linear sequence in which we must read and write of them but massively and all at once, as an interwoven whole experience. The meanings of certain events and decisions are becoming clear only now; many of the social consequences, however, were predictable at the time they occurred.

The Governor Acts

Governor Hugh Carey was the chief executive of New York in August 1978 when Health Commissioner Robert Whalen catalyzed apparently isolated events in the western corner of the state into a dynamic, far-reaching whole. The commissioner had declared a health emergency. Stating that the Love Canal area in Niagara Falls was a hazard to health, he had ordered that a huge remedial construction project be undertaken to correct part of the problem, had ordered continuing health and environmental studies, and had recommended that pregnant women and children under two, and their families, move from the region—temporarily, but as soon as possible.

Commissioner Whalen, working in Albany, far from the Love Canal scene, and relying on reports, may have underestimated the impact his pronouncements would have on the populace. Equally likely, he may not have considered people's reactions as part of the set of problems he and his department must consider. A high-level official, in response to an informal query, replied that the health department professionals were *scientists,* who did not worry about people's reactions to cautionary statements and recommended actions. They dealt with numbers—with data on physical conditions—and only with these. Political and social matters, the official stressed, were extraneous to the DOH work. The issues of how pregnant women and children would move to safe places, how people might feel about recommendations from the state health commissioner not to go into their basements and not to eat food grown in their gardens, were not seen as the responsibility of scientists.

Political and social matters were of great importance to others, however. Love Canal residents were keenly aware of what they regarded as insensitivity to them as whole, feeling, thinking social beings when the health commissioner wrote his order without the residents' participation and then announced it in a city three hundred miles from the scene. He sent copies of his order to fourteen offices and "interested parties," but not one Love Canal resident was on the list. The affected people, learning their fate from newspaper, radio, and television reports, resented what they perceived as high-handed treatment.

Whenever Governor Carey learned about the environmental and health problems at Love Canal, and whatever he had done prior to the health commissioner's directives, the governor was not yet visible as a presence responding to the widely publicized events occuring in the heavily populated western section of his state. Despite preliminary investigations, the extent of the physical problems was not totally clear, but there were strong pressures on the governor to attend publicly to Love Canal at this time.

No one had precise information on the contents of Love Canal, the migration patterns of the chemical wastes, or the exact effects of long-term human exposure to a mix of chemicals. The chemical leachates, however, were discernible in some Love Canal homes and yards, and serious effects could be anticipated beyond the increased miscarriage rates already calculated for that region by the health department.

By mid-1977, the EPA had reported that Love Canal leachates contained, among other things, PCBs and hexachlorophene, substances known to be highly toxic to living beings (Moriarty 10/18/77; MacDonald 9/20/78). In the summer of 1978, the department of health identified more than eighty chemicals in the Love Canal landfill, ten of them "known or suspected [to cause] cancerous growth in laboratory animals, and one—benzene . . . a well-established carcinogen" (NYDOH 9/78, p. 12). These words appeared in a report the department of health prepared for Governor Carey and the New York State Legislature to justify their request for a half-million dollars to begin further health and environmental studies at Love Canal. Entitled *Love Canal: Public Health Time Bomb,* this report was widely publicized and freely distributed soon after Dr. Whalen's order of August 2. Along with data analyses describing elevated miscarriage and birth defect rates at Love Canal, ten compounds were listed, with some indication of their toxicity to living beings (table 4-1).

Table 4-1
Some of the More Important Chemicals Identified at Love Canal and the Human Biologic Hazards Associated with Them

Compound	*Acute Effects*	*Chronic Effects*
Benzene	Narcosis Skin irritant	Acute leukemia Aplastic anemia Pancytopenia Chronic lymphatic leukemia Lymphomas (probable)
Toluene	Narcosis (more powerful than benezene)	Anemia (possible) Leukopenia (possible)
Benzoic acid	Skin irritant	
Lindane	Convulsions High white-cell counts	
Trichloroethylene	Central nervous depression Skin irritant Liver damage	Paralysis of fingers Respiratory and cardiac arrest Visual defects Deafness
Dibromoethane	Skin irritant	
Benzaldehydes	Allergen	
Methylene chloride	Anesthesia (increased carboxy hemoglobin)	Respiratory distress Death
Carbon tetrachloride	Narcosis Hepatitis Renal damage	Liver tumors (possible)
Chloroform	Central nervous narcosis Skin irritant Respiratory irritant Gastrointestinal symptoms	

Source: Reprinted from: New York Department of Health, *Love Canal: Public Health Time Bomb*, September 1978, p. 12.

As citizens of the state, Love Canal residents were entitled to some basic health and welfare protection, and they knew it. From many residents' viewpoint, there had been signs of trouble for years; the Hooker Corporation and lower-level government agencies had passed the buck until this point, and now the citizens were demanding what they believed was due them.

Furthermore, the whole affair was hot news all over the world. Not only were cameras focused on the tearful faces of outraged people of all ages and conditions at Love Canal, but the governor's every move was under scrutiny, for 1978 was a gubernatorial election year. Any decisions the governor made—to initiate or implement solutions to the Love Canal problems—would be newsworthy, resulting in advantage or disadvantage for his election campaign (Allan 8/15/78). Resources had to be mustered swiftly and utilized decisively, for the Democratic party primary election was only six weeks away.

At that point, early in August 1978, Governor Carey was facing a primary challenge from his lieutenant governor, Mary Ann Krupsak, and the early polls suggested his reelection was not a sure thing (Logan 7/30/79). The problems his aides had already encountered in dealing with the Love Canal problem had been broadcast in all the newspapers and on television. During the stormy session of August 3, 1978, when the governor's aides conveyed his deep concern to the audience, the aides were soundly booed. Their explanations of his efforts on their behalf, including his appeal to the president for federal disaster relief (Governor's Office, 8/3/78) and his plan to meet with federal and state officials the next week, were to no avail.

The health commissioner's August 2 recommendation that selected people move from the area was a matter of prime importance to the majority of Love Canal residents. Many reasoned that, if pregnant women and young children were to move for health reasons, then living near the canal might seriously endanger anyone. The residents also were alarmed about the devaluation of homes and neighborhood that the recommendations created.

The anger people directed toward public officials who spoke of no financial resources was in marked contrast to their joyous expressions of gratitude for promises of help. Several hundred assembled residents accorded Congressman John LaFalce a standing ovation on Friday evening, August 4 when he addressed the meeting at which the Love Canal Homeowners Association was established. He brought the news that President Jimmy Carter was backing an amendment to the Resource Conservation and Recovery Act (RCRA) to use $4 million for remedies at the dump site.

While the Governor's aides were cautioning about a lack of resources, the Hooker Chemical Corporation spokesman, though continuing to emphasize the company's lack of legal responsibility, announced the company's willingness to lend technical skills to the government to help correct the situation (*Courier Express* 8/4/78). Furthermore, Mary Ann Krupsak, the governor's primary election opponent, continued her scathing comments on the governor's inattention to the problem and called for a full evacuation of the people (*Niagara Gazette,* 8/5/78b).

The open messages the governor was receiving were quite clear: no matter how hard he might be working to marshal resources behind the scenes, it appeared to the public that there was little or no response from his office; meanwhile, others were criticizing him, responding to the people's needs, and looking good—facts widely reported and commented upon. The time to act had drawn nigh.

One of the governor's first acts was to create a multiagency administrative body, formally entitled the Governor's Love Canal Inter-Agency Task Force, to address the Love Canal problem. The list of agencies in the task force started with the departments of health and environmental conservation, the first state agencies involved in Love Canal work. By early August, these two agencies and seven others comprised the state task force. It was organized at two levels, with the policy decision makers located in Albany, close to the governor but far from the day-to-day operations of the task force workers on site in Niagara Falls and far from the daily plight of the residents.

The head of the task force was William Hennessy, the commissioner of transportation, while Mike Cuddy, a New York Department of Transportation (DOT) career officer, headed the on-site operations. The DOT personnel were considered well suited to achieve the major task force goal of overseeing the massive construction project, since they had experience in overseeing highway construction projects and in working with citizens in connection with these projects. The majority of the residents became familiar with the personnel and the work of the DOT and of the departments of health and social services, for these three had the greatest amount of face-to-face contact with residents. The remaining agencies played technical roles, remained in the background, and were not as visible to most neighborhood residents.[1]

The state of New York was now very visibly present, and the next question was exactly what could and would be done. Within a few days, there were favorable signs that federal support would be forthcoming to aid state efforts to end the nasty problems at Love Canal once and for all.

Signs of Support: Maybe Something *Is* Blowing in the Wind

At the behest of Congressman LaFalce (LaFalce 12/24/78), the Federal Disaster Assistance Administration (FDAA) administrator, William Wilcox, arrived in Niagara Falls on Saturday, August 5. During his inspection, Wilcox was accompanied by aides to Congressman LaFalce and Governor Carey and by officers of the Homeowners Association and other residents. Wilcox spoke with the residents, assuring the Homeowners Association president that he wanted citizen participation directed to him, for he felt that cutting red tape was very important to enable swift action (Cerrillo and Gibbs 10/10/78).

He saw leaking basements and was photographed while spontaneously frowning and holding his nose, and he listened when some residents told him, emotionally, of their desires to get themselves and their children out of the area (Seal 8/6/78). He left the residents, and evidently the government officials, with what seemed to be assurances that federal aid would be arriving. He was quoted: "My personal impression is that this is a very troublesome site from the public health standpoint and I feel confident that some federal aid will be made available" (Seal 8/6/78). He also stated that there was nothing in the Disaster Relief Act Amendment of 1974 "that would ordinarily preclude" a situation like Love Canal. He said it would come under the category of "other catastrophes" (Seal 8/6/78; see Disaster Relief Act of 1974).

On Monday morning, August 7, John LaFalce read the jubilant local headlines to a White House aide, to make sure that President Carter "got the message." The White House, commented LaFalce, was well aware of the political danger of refusing aid in an emergency (LaFalce 12/24/78). That day, the Senate approved, by a voice vote, that "Federal aid should be forthcoming" for Love Canal (NYDOH 9/78, p. 25). That very evening President Carter declared that an emergency existed at Love Canal. The welcome declaration made the area eligible for the help of FDAA funding and agencies to "save lives, protect property, public health and safety, or avert the threat of a disaster." (McNeil 8/8/78). No specific funds were described at that time, but the regional FDAA official, Norman Steinlauf, was appointed to coordinate efforts. The other source of federal funds, the $4 million for a RCRA demonstration project, also seemed to be moving along nicely. Senators Daniel Moynihan and Jacob Javits were prepared to act. Their amendment had been reported as having the approval of the powerful director of the Office of Management and Budget and of the EPA, which administers the RCRA legislation. It seemed that the federal sources were finally working together to release essential funds for both the construction work and the relocation of residents (Lynch 8/5/78).

"We'll Buy Your Houses"

On Monday evening, August 7, Governor Carey visited Love Canal for the first time, arriving about four hours after President Carter's emergency declaration. Although he was two hours late for a meeting originally called for eight o'clock the governor was the crowd's darling that night—greeted by enthusiastic cheers and applause from the Love Canal families crowding the 99th Street School auditorium.

Governor Carey pledged that state emergency aid would be available (S. Johnston 8/8/78). He then gave the now hushed, expectant crowd, the "fantastic news" that the state was going to buy the homes bordering the Canal! Prior to the public meeting, he had met with some members of his newly organized Love Canal Inter-Agency Task Force and with some residents, including Lois Gibbs. Gibbs, president and spokesperson for the new Homeowners Association, had relayed the residents' desire to be relocated and their fears of being in the area when the construction started. One observer of the public meeting remembers that "someone shouted, 'Governor, I live right across the street from the homes next to the canal. My house is contaminated, too,' and the governor said that the state would buy those, too. Then someone from 102nd Street spoke up and he [the Governor] said 'No, that's it. There is going to be a line. We'll put a fence up.' "

When the residents asked the governor whether they should take their household goods with them when they left their homes, he said, referring to the possibilities of contamination, "I don't have to be a doctor to answer that. Leave it behind." (Perrault 8/8/78). Tom Heisner, who had been helping to organize the tax strike, told the governor, with the obvious assent of the assembled crowd, that, if Governor Carey did everything he said he would, they would be happy (Perrault 8/8/78).

Now the bureaucratic staffs had to deal with the elected official's promises—not weeks hence, after an election, but immediately. The good feelings the people received from the governor's evening visit evaporated in the light of day, when state officials on site told the residents from the far sides of 97th and 99th Streets that Governor Carey had *meant* that they would be relocated *only* if they could show they had health disorders directly connected with materials in Love Canal. This interpretation stirred these residents to anger, for the air readings some of them had received from the state showed higher figures than did those of some neighbors who were slated to move. The governor's statements had not sounded that way to them when they sat in that audience the night before. When they saw the preparations for relocation of families whose backyards touched the canal fields—fearing that the governor's promise was to be broken—groups of people gathered and went to the 99th Street School to tell the state task force workers exactly how they felt. That day, a brave state division of property official faced a crowd of a hundred angry residents, telling them he had no orders to relocate them (Shribman 8/9/78). Smaller groups of residents continued to badger the state officials in Niagara Falls with questions and angry outbursts.

Within a few days, the governor made it clear that he, not the on-site officials, "was in charge of the state machinery." He wanted to work *with* the residents, not against them. The residents were much relieved when the governor issued a statement that the state definitely would move all families living on both sides of 97th and 99th Streets and on the section of Colvin Boulevard that lay between—a total of 239 families—if they wished to leave. For these inner-ring families, the panic was over for the moment (Brydges 8/10/78). For the residents of the other streets, however, starting a few short yards from the chosen group's homes, their time of mental anguish was to continue to grow worse.

The Remedial Construction Project

The very first solution proposed for the Love Canal problem was to stop the leaching (the migration of chemical-laden water), to prevent future leaching, and to cover the canal up tightly. For about a year previously, the biggest difficulties with that solution had seemed to be straightforward: finding the right people to do the right job and finding the money to pay for it. It had been fairly simple to find engineering firms to assess the situation and prepare construction plans. When Commissioner Whalen issued his order on August 2, the city of Niagara Falls and the Niagara County Board of Health were ordered to implement the already-prepared Conestoga-Rovers plan for the southern portion of the canal, subject to the approval of the plans by the department of environmental conservation.

The Conestoga-Rovers multimillion-dollar remedial construction plan called for the digging of trenches and the installation of a tile drainage system to divert the leachates to a trench system, from which they would be pumped out of wells to a holding tank, then treated to remove organic toxic wastes, and finally emptied into the city's sanitary sewer system. The canal itself was to be covered with an eight-foot-thick compressed-clay cap to prevent rain and groundwater from entering it. Although the project plan contained recommendations for further studies to discover whether the bedrock and underground waters had been affected, the tone of the proposal was optimistic. The plan contained the prediction that the proposed construction would be so effective that homes across the street from those directly abutting the canal would become habitable upon completion of the work (Conestoga-Rovers 8/78; Shribman 8/27/78).

This prediction was remarkable, since the extent of migration of the leachates into the surrounding areas was unknown then and still remains unknown. There were many avenues of migration for the chemicals. Although Love Canal was lined with clay, there were dessication cracks in the lining and in the ground clays that stretched away from the canal and lay below five to twelve feet of surface soils, which were composed of more permeable silty clay loams. The entire area was described as rich with a "maze of drainageways," including creeks and numerous shallow swales (natural drainage ditches), and even a trench leading to the Niagara River. There were utility lines traveling in irregular patterns underground, storm and sanitary sewers, and drainage pipes that had been installed at various times, some in gravel trenches (Ebert 3/79). The construction of the school and then homes, with work going on for years, may have affected the flow of materials from the canal. The soil had been disturbed frequently, possibly moved to and from the surface of the canal and its east and west extensions of land. The streets cut through the area in the late 1950s and early 1960s were embedded in large stones, which served as channels directing chemicals toward the streets they crossed (Lester 7/25/79). When the expressway was built between the canal and the Niagara River to the south, the enormous foundations of that roadway may well have impeded drainage from the canal, thus forcing more materials to travel paths of creeks and trenches and other routes (Ebert 3/79).

In light of all the pathways for migration, it is understandable that some of the homes later found to be contaminated were at a considerable distance from the rectangle of the canal, which lay embedded in a 200-foot-wide strip of land. Later, the health department asserted an alternative hypothesis—that contamination in outlying areas was caused by the transportation of contaminated earth used for landfill (NYDOH 4/81). What is puzzling, however, is that the engineers' project statement was so confident in its promises amid such uncertainties as to the location of the leachates. We may only guess at the reasons for their confident tone. Engineers, accustomed to working with natural forces, routinely compensate for the unexpected. They deal, however, with a relatively small number of physical variables, whose behaviors are highly predictable. The uncertainties at Love Canal may have seemed no greater to them than those in any large project. There also may have been social reasons for their confident statements. In order to win contracts for expensive projects, one must appear confident about one's own proposals. When funds come from public sources, it is sometimes necessary to oversell, promising more than can be delivered reasonably (Levine and Levine 1975). Studies of role playing and dissonance reduction show that the forceful, positive presentation of one's viewpoint can lead to an increased subjective sense of confidence in the position one is publicly espousing. In this instance, not only the engineers responsible for the project had to defend it; some government officials, having committed themselves to this plan, also defended the project against questions and criticisms from the residents in public meetings where they were told about the construction proposal. Thus, the engineers and government officials may have bolstered each other, making the predictions more rigidly optimistic the more they were publicly criticized (see Janis 1972).

The city and state officials had committed themselves to the Conestoga-Rovers plan early on, rather than allowing for the possibility of alternative procedures in the event that criticisms of the plan's underlying assumptions or mode of procedure proved to be correct. Allowing for alternative bids for the project may not have been essential legally in the declared health emergency. It would have been scientifically appropriate to consider alternatives, however, and it undoubtedly would have made good sense to the residents, who fully understood the extreme difficulties of locating liquids that had been moving underground for up to thirty years. Since only one plan was under consideration, however, and since that plan was mandated in the health commissioner's order of August 2, the possibility of alternatives was not among the "thinkable thoughts" at that time. As far as the authorities were concerned, the Conestoga-Rovers plan was *the* correct one, and that was that.

One early critic of the construction plan was Wayne Hadley. He argued that, first, the extent of the chemical migration should be assessed or at least estimated carefully. Then boundaries of a potentially unsafe zone should be set—well beyond the estimated area—both in order to maximize the safety of the people, who should be moved from the unsafe zone, and to maximize the possibilities of doing remedial construction where it would be most effective. His idea of making careful estimates of the extent of the migration of chemicals before investing in a multi-million-dollar construction project was not heeded. The Conestoga-Rovers plan called for a huge, rectangular cover to be neatly placed over an area defined by the 200-foot-wide strip of land within which the canal was embedded, and so it was to be. Once the construction project work began in October 1978, environmental testing continued, but no changes were made in the area to be covered.

The comments of other professionally experienced people regarding the construction plan were less than positive. The president of Chem-Trol Pollution Control Service withdrew his company from the bidding on the remedial construction after he examined the proposed plans, saying that the plans contained "too many ifs, ands, and buts" (Perrault 8/18/78). Professor Charles V. Ebert, a geographer and soils expert at the State University of New York at Buffalo, had volunteered his services to all interested parties. He pointed out some inherent weaknesses in the plan—related not to the removal and containment of toxic wastes from the canal but to the chemicals that had already migrated beyond it. Ebert urged that studies of the underlying soil, to the bedrock in all directions, should commence immediately. He warned that long-term monitoring of the completed construction work would be indispensable for determining the effectiveness of the construction project (Ebert 9/21/78; Brown 9/27/78). Another of the residents' consultants later said that the system might very well drain water from a distance but still leave chemicals clinging to the soil, since chemicals do not necessarily behave like water (Lester 7/25/79).

Some residents were skeptical about the construction plans from the very beginning, raising questions about the project's effectiveness privately and in public encounters with city and state officials and with Frank Rovers, whose company produced the plans. The questions were not always raised in the most polite terms (Porter 5/17/79). Many Love Canal people were employed in construction or in the local water-treatment plant, and they had seen experts' plans sometimes go awry in practice. They questioned details of the proposed construction, and they were not in a mood to admire its engineering elegance. They angrily attacked such reassuring statements as the assertion that the pipes or the filters in the trench system would never become blocked.

The public sessions where such questions were asked were not pleasant. At sessions in the spring and summer of 1978, and even later in the winter and early spring of 1979, those in charge of the meetings seemed to take refuge in being very firm, in acting unconcerned about contradictions, and in never admitting the uncertainty that was, of course, inherent in the situation. Far from reassuring residents, such firm behavior made them think that the authorities were not to be trusted with the safety of Love Canal families. It was in part that disquieting feeling from the beginning that provided some of the impetus to the citizens to organize. The official response then was to develop a safety plan, which we will discuss in the next sections.

First, however, we shall consider how well founded the residents' criticisms of the remedial construction project turned out to be. By late December 1978, two months after the project began, it was necessary to modify the original construction plans in order to dig lateral trenches at right angles to those parallel to the canal, in an attempt to increase the flow of leachates into the major draining system (Kostoff 12/30/78). In mid-July 1979, the tiles in the southern section (where the project commenced) clogged seriously, necessitating immediate corrections to the system, which took weeks to accomplish.

Two years after the project began, a consulting firm prepared an evaluation of the project for the New York Department of Transportation. The summary stated that the remedial construction plan "was designed to contain only wastes deposited in the canal itself" and that "the clay cap had reduced chemical evaporation from the Love Canal and has minimized rainwater infiltration" (Clement 8/20/80). Despite Dr. Ebert's early admonitions, no monitoring system was installed to check the ability of the gigantic construction project to intercept the toxic wastes and prevent their outward migration. The reported reason, given in June 1980, was that conditions imposed by the state department of environmental conservation lawyers drove the costs beyond the funds allocated for the purpose (MacClennan 6/12/80). In November 1980, the Environmental Protection Agency reported that the treatment plant at the canal site was removing high levels of chemical contaminants from holding tanks, where materials from the perimeter of the canal were collected. The EPA, it was reported, planned "to determine if the overall project is successful in stemming the flow of chemicals into the canal neighborhood" (MacClennan 11/5/80). The consultant's report, however, pointed out that the remedial construction project "was not designed to reduce contamination levels beyond the immediate canal property. For this reason, the remedial construction plan is only a first-step solution to the contamination problems of the area" (Clement 8/20/80, p. 1). The residents, watching and reading about the later corrections, evaluations, and plans for state investigations of the work (Westmoore 10/18/80; *Niagara Gazette* 11/17/80), were to feel more than ever that some of their predictions had proved at least as good as the expert's early opinions, and that their concerns had been well justified.

Developing a Safety Plan; Developing a Group

The development of a safety plan to protect residents during the proposed remedial construction project was an important rallying point for the newly organized residents. Forcing the state to provide them with an acceptable plan was a short-term goal with very real meaning to the worried people. Their participation in various key episodes while the plan was developed swiftly plunged the citizen leaders into some of the realities of bargaining and maneuvering with governmental officials.

State and city officials had conducted public meetings with the residents as early as May 1978 not only to describe the proposed epidemiological and environmental studies but also to tell residents about the remedial construction plans proposed by the Conestoga-Rovers firm. During the meetings, the residents criticized the plans and deplored the level of information they received. The officials believed they were offering plans for tangible help, and, surprised at the anger expressed, they became somewhat shaken in conducting meetings where dozens of people asked probing questions that the officials were unprepared or seemed unwilling to answer.

Those early public meetings provided the context in which the residents began to learn that they shared common problems and in which they also began to develop the deep understanding that they would have to depend on themselves and one another. In short, a *group* was emerging, composed of "the regular people who live here, not the ones here working for the government." The boundaries of that *group* became sharply defined in the people's minds at the first meeting of the Love Canal Homeowners Association on August 4, the night following the stormy session at which Commission Whalen had faced the angry crowd chanting "We Want Out!"

On the night of August 4, the local firehall was packed with several hundred people. In addition to cheering Congressman LaFalce when he arrived with his positive report about funding, the large, excited assemblage conducted important formal business. One of the first questions they settled was what constituted the Love Canal neighborhood. The vast majority voted to include their homes and declared that the appropriate region was from 93rd to 103rd Streets, from Buffalo Avenue to Colvin Boulevard. That large square encompassed 789 single homes, some 250 rental units for low-income people, some senior-citizen housing, a few commercial properties, and, in the very center, Love Canal itself, with the 99th Street School on top of it. From that night on, relations between the government and the residents were strongly influenced by the fact that the residents were organized and that they seemed to present a fairly unified force.

One of the Love Canal Homeowners Association's first acts was to engage Richard Lippes, an attorney interested in environmental issues, as the association's legal representative. In one of his first public statements, Lippes commented on the remedial construction plans, pointing out that the residents feared the release of additional poisons in the course of the work. The plans, he said, addressed the problems of the physical site but not the problems of people living nearby (Rosen 8/6/78). Lippes's comments heralded the dispute that soon arose about the lack of safety features to protect the residents in case the work at the abandoned chemical-waste-disposal site released toxic chemical fumes or caused an explosion or a fire while hundreds of people were going about their daily lives, some less than a hundred yards away from the work site.

A few days later, Lois Gibbs requested and received a copy of the safety plans for the proposed construction work from the task force safety officer. She noted that the work plan assumed that there would be digging into the canal itself and included precautions to use nonsparking equipment, with warnings about the potential explosiveness of the phosphorous rocks. Gibbs complained to the safety officer that, although the safety plans discussed precautions for workers, residents living near the proposed work site were ignored. She asserted that the residents would be very upset at the prospect of someone digging directly into the canal. She knew that some residents' fears were based on their witnesssing, reading, and hearing about explosions and fires on what they now knew was Love Canal. When the Calspan Corporation's 1977 report was revived and publicized at this time, the residents were reminded that Calspan urged that digging proceed with "archeological caution" (Desmond 8/10/78).

During those days, with their swiftly moving events, Lois Gibbs spoke frequently with the governor's aides. She repeatedly told them that, if she were to show the residents the safety plan and relate to them the slogan that the aides recited to *her*—"a good on-site plan is a good off-site plan"—there *would* be an explosion, but it would come from the people. Furthermore, she hinted strongly, it would occur on camera. By now, the public meetings were no longer occasions where residents simply received information and spontaneously expressed their emotions, as individuals caught up in a frightening situation. The emotions people continued to express publicly were very real and deeply felt. Cameramen took countless news photographs and film strips of distressed residents eloquently expressing their plight to officials, who were maintaining their dignity with outwardly calm demeanors. Those recorded moments shaped the spontaneous events into marketable packages containing a sort of "goods" that was useful for news producers. The residents, however, quickly became more than the passive subjects whose recorded images were used by others. They also began to use these captured moments by transforming them into weapons; that is, the organized residents could threaten the officials and agencies they represented with embarrassing public exposure, because their problems and reactions were newsworthy. Particularly in early August 1978, the ongoing political campaign, with the primaries still ahead, provided a context in which the threat of publicizing citizens' reactions was potent.

On August 9, Lois Gibbs was invited to a hastily assembled high-level meeting in Washington, where a variety of government officials discussed the Love Canal problem. She took the opportunity afforded her when William Wilcox, head of the Federal Disaster Assistance Administration (FDAA), met her at the airport. (The FDAA was the chief agency from which the state hoped to receive funds.) During the taxicab ride, she showed him the safety plan for the project, which was supposed to start on August 15, well before any families would be out of their homes. Gibbs told Wilcox that the residents would be "hysterical at the thought of having canal contents disturbed" while they were living nearby, and he replied that a new safety plan would be recommended by August 10.

On August 11, she awaited the plan rather anxiously, for she wanted to discuss it with the board of directors of the newly formed Love Canal Homeowners Association before the first general meeting of the association to be held on August 14. To her chagrin, the task force released the safety plan to the press in time for the Saturday, August 12, editions before it was given to her. This action made her and the board of directors feel that the citizens organization was neither recognized nor treated with respect by the state—her discussion with Wilcox and her attendance at the Washington meeting notwithstanding.

The starting date for the construction was now postponed from August 15 to August 21, but the work did not start until October 10, for two major reasons. Not only did it take longer than anticipated to locate satisfactory housing and to move residents from their homes next to the canal, but the safety plan was angrily denounced from the time the residents read about it in the papers.

The on-site safety plan was developed by the Hooker Corporation as part of its effort to provide help and technical assistance when possible (*Niagara Gazette* 8/14/78)—a bit of corporate good citizenship that may have been less than reassuring for the residents at that time. The plan included "paramilitary precautions, with an oxygen unit, detoxification showers, and disposable work outfits . . . backed up by standby police, fire and closed circuit communication." Equipment was available to monitor the air and to contain chemical materials escaping during the construction process (Shribman and MacClennan 8/12/78). At this time, people whose backyards bordered the canal area on 97th and 99th Streets and faced it on the strip of Colvin Boulevard between the two streets learned that they lived on the "first ring" or "ring one." The "second ring" described homes across the street from the ring-one homes on 97th and 99th Streets. The two rings together comprised the "inner ring," and all the rest, from 93rd to 103rd Streets, were the "outer rings." The designations served the purpose of identifying the target area. The connotations were not lost on residents, who were startled by their transformation from individuals and families to labeled units in a mass to be processed in some quasi-military fashion—if, in fact, they were to be protected at all from the target-zone dangers.

In many residents' view, there were severe inadequacies in the safety plan and, in fact, in the whole remedial construction approach. City and state officials promised that no construction work would start until all the first-ring families were moved out, but that did not address the outer-ring dwellers' fears. (Brydges 8/15/78). The people living in the outer rings were learning that toxic chemicals were being found every day in areas well beyond what originally had been described as the perimeter of the canal. (MacClennan and Shribman 8/11/78). Furthermore, they feared that chemical-warfare materials had been buried in the area and were about to be released—the army's denial notwithstanding. The publicized information about the geographic extent and nature of the materials in the canal undercut the soothing reassurances offered by government officials. Not only was there no safety plan for them, but, from many residents' perspective the massive construction plan simply would not eliminate or even mitigate contamination in the extensive area beyond rings one and two, and no other plans were forthcoming for an extended cleanup of the area (Shribman and MacClennan 8/16/78). Perhaps most of all, the people living in the outer rings were apprehensive that, once the construction work started—but even more once it was complete—they would simply be left in their dangerous, now unsellable homes, forced to take huge financial losses if they ever were to get out.

Lois Gibbs, exercising the power of withholding a resource, now refused to call the general meeting of the Homeowners Association that had been scheduled for August 14 to discuss the new safety plan. It seemed to her and to the association's board of directors to be only half a safety plan, with no procedures for the safety of the residents, no plan to temporarily relocate people living outside the perimeter of the canal, and not even recognition that residents might have to be evacuated in the event of a sudden explosion or release of toxic vapors (Shribman and MacClennan 8/14/78). Matters came to a standstill. It took Governor Carey's personal visits and words of assurance to get things moving again.

Learning that There Is a Political Game

The governor continued to show his solicitude for the Love Canal people during this preprimary period. Taking advantage of an engagement in Buffalo, he dropped by the 99th Street School in Niagara Falls on August 15 to assert that he meant to keep all promises he had made to the residents. Michael Cuddy, the on-site director, reportedly assured him that communications between the interagency task force and the residents would improve. There was now a liaison person, who was to be the governor's special emissary, working with the residents. Also, resident's representatives would be invited to weekly task force meetings. The most positive sign that residents would be included as part of the team, however, was the office in the 99th Street School that was given to the Love Canal Homeowners Association upon their request and equipped with secondhand but adequate furnishings and two telephone lines, all underwritten by the state. This former teacher's lounge was to be a crucial asset for the growing organization, a headquarters where information was collected, contacts were made, and many citizen activities took place. Later, when a second organization arose, representing low income renters, they were provided similar facilities.

The governor gladdened the residents' hearts when he was somewhat abrupt with Mayor Michael O'Laughlin. He told the mayor publicly that nothing would proceed until the residents' safety was assured, despite the mayor's concerns that a delay in starting the work would jeopardize the city's chances of procuring federal funding (Spencer 8/18/78). The governor "couldn't believe" that the safety plan had no "provision for the safety of residents living nearby the construction area." He promised that the construction work simply would not proceed until the nearby residents were out of the area (Delmonte 8/16/78). He told Lois Gibbs, who quickly relayed the message to the Homeowners Association, that "not the heel of the boot of a worker" would be placed on the construction site until the association's approval of safety measures was secured.

The association officers, following Wayne Hadley's lead, had complained loudly about the lack of citizen "input" from the beginning. Now they were invited, for the first time, to attend a meeting of the interagency task force, to be held August 18. William Hennessy, the commissioner of the department of transportation, who was the state task force chairman, was to chair the meeting. A number of other state and city officials were to be present, as well, to discuss another revision of the safety plan.

Since it was only two days after Lois Gibbs's public refusal to call a Homeowners Association meeting to discuss the first revised safety plan, the timing of the governor's visit on August 15, suggested that he was at least willing to placate the residents. On August 17, however, an incident happened that first puzzled and then angered Gibbs and Cerrillo, the association's leaders. The night before they were to attend their first task force meeting to discuss the revised safety plan, with the board of directors' approval, they suggested to the governor's aide who was acting as a liaison that the task force meeting be postponed, because they wanted adequate time to review the safety plan before the meeting. According to Cerrillo and Gibbs, this aide "had been the nicest, best friend before this," but now she forcefully dissuaded them from this course. She knew that cancellation of a meeting of many state and city officials would be inconvenient and awkward to explain, since part of her task was to keep communications and working relations operating between citizens and the state. She emphatically stated that they could not cancel a meeting and make a scene that would embarrass the governor. Gibbs and Cerrillo tried to tell her they did not want to embarrass the governor but simply wanted the meeting postponed to allow them time to examine the plans more carefully. In the course of the discussion, Cerrillo and Gibbs later recalled, she heatedly proclaimed that the perimeters were established because it was a political decision, *not* a health hazard, and she told the two women to get the latter idea right out of their heads. She told them that neither she nor the governor had put the chemicals into the ground, that the governor was the only resource they had, and that the task force could be swiftly withdrawn, never to reappear, if the residents' representatives were obstinately uncooperative (Cerrillo and Gibbs 10/10/78).

Intimidated by this threat, Gibbs and Cerrillo attended the meeting on the following day with thirty state and city officials and a handful of other residents. At the meeting, task force chairman William Hennessy announced that the state was prepared to proceed with necessary actions to initiate the construction work, despite the residents' reservations about the lack of clear plans for their safety. Commissioner Hennessy said that, although he truly preferred to work *with* the residents and to find, rather than impose, solutions, he had the power and the authority under executive law and the health department's emergency declaration to take all steps necessary to arrest the flow of toxic chemical wastes. In response, Lois Gibbs took the safety plan and tossed it into the center of the huge table, remarking that it was a useless document. The gesture was ignored by everyone present except one television reporter, who showed the brief act later in a local news bulletin.

The Homeowners Association attorney, Richard Lippes, responded to Hennessy by pointing out that imposed solutions would be subject to legal delays. At the end of the meeting, the commissioner promised to postpone the start of the remedial construction for several days (Wells 8/18/78). Very shortly, the task force announced that the construction would be delayed until the inner-ring people were evacuated.

The episode taught the citizen leaders a few lessons about the uses, sources, and limits of power. Both William Wilcox (FDAA chief) and Governor Carey had promised, or strongly implied, that the massive construction project would be delayed until the residents approved plans for their safety. Those promises seemed a solid recognition of the importance of citizen participation, a central concept in our democracy. That apparent delegation of authority gave the citizens some control in the situation, for the beginning date of the remedial construction project was thought to influence whether the Love Canal situation would be funded as an emergency by the federal authorities. There were limits, however, because the state officials were indeed anxious to start the work—not only for purposes of funding but also so that it would be well launched before the bitter winter weather began. They were willing to use their full mandated power to carry out what they deemed necessary steps, with or without the residents' approval. They, too, were limited, however. When two of the governor's aides scolded Lois Gibbs the next day for embarrassing the governor when she literally discarded the safety plan, the citizen representatives learned again that, at least in this preelection period, the state officials would pay heed to people who led a large, vociferous, organized group and who retained the ability to influence voters by actions and words that were embarrassing to elected and appointed officials up to the highest levels. The young women also saw, by their attorney's cool conduct, that technical weapons could be counterposed to actions on the state's part by a person who knew precisely what the issues and the rights were in each incident. These lessons, drawn from a complex episode, guided the leaders of the citizens group.

They were learning something about power at a level close to their own experiences and were beginning to sense that the political game played in a larger context was affecting them. They did not yet know how that game operated, but Lois Gibbs began to try to play, too. She wrote a forceful statement criticizing the revised safety plan, the task force, and, by implication, the governor. When she spoke with task force staff members, she threatened to release her statement to the press, and she hinted about it to the journalists, who then reinforced her threats when they asked the task force staff for comments.

There was to be a large public meeting of residents on August 24 to discuss the twice revised, still unaccepted plan. The evening before, Governor Carey attended the ballet in nearby Lewiston, New York, and then held an impromptu meeting with Love Canal residents at the 99th Street School. The governor listened attentively to descriptions of sick children, frightened pregnant women, and other people who felt particularly endangered by the construction work. He responded to their fears of remaining in the area during the construction, safety plan or no, by telling the assembled residents, "Where there's a medical diagnosis that there can be an impairment to an individual's health, we will tend to that person" (MacClennan and Shribman 8/24/78). Although the governor indicated the state was not ready to relocate families in the outer rings unless it was shown that a specific hazard existed in individual homes (Brydges 8/24/78), his words were reassuring. Lois Gibbs did not release her statement. She and others who heard him did not doubt that the governor fully intended to take care of people with health problems and with demonstrable chemical presence in or near their homes. She remembers: "I flew down the hall and told the others that everything was going to be okay. We weren't going to be abandoned. I was so happy. What a naive fool I was! I thought we had really gotten through to the governor."

Lois Gibbs *was* naive at that point. Her forceful statement notwithstanding, it did not occur to her that firm policies already in place by mid-August might not include any further considerations of relocating residents, whether on a permanent basis or temporarily during the period of the construction work. There were facts involved beyond the local scene, stemming from that higher-level political context she was now sensing. Before turning to the major reason that policies were set by this point in August, we will describe the safety plan that was finally implemented, even though it was never formally accepted by the skeptical residents.

"When You Can Hear that Whistle Blow, It's Too Late"

At public meetings on August 24 and 31, 1978, a contingency evacuation plan was distributed to the Love Canal audience. The diagram illustrating the procedures to be undertaken " in the most unlikely event of such need arising in the course of construction" seemed extremely complicated at first. Essentially, it called for fifty-three school buses to be stationed, with engines running, on the streets in the third, fourth, and fifth rings—that is, from the soon-to-be-fenced-off boundaries of the evacuated streets through an area bounded on the west and east by 93rd and 103rd Streets and to the north and south by Colvin and Frontier Boulevards. Some 800 families lived within that area, in addition to the 239 whose homes could be purchased. There were to be standby police and ambulances. If an explosion, fire, or release of noxious gases occurred, the police were to drive through the neighborhoods, sounding an alert with bullhorns (not with sirens, because the chemical companies in Niagara Falls use siren warnings). Upon hearing the bullhorns, people were to pick up suitcases they were to have packed and waiting by their front doors, collect their children, and proceed quickly and in orderly fashion to the buses. They were to leave all possessions behind, and the family pets as well. Once on the buses, they would be taken to the Niagara Falls Convention Center, there to be reunited with children who might have been in school and with other family members.

At the first meeting, August 24, people puzzled over the contingency evacuation diagrams, mumbling quietly to each other until the meeting was called to order. Then they began to ask questions. Some hissed and hooted when they heard that the buses would cost over $100 a day each, about

$30,000 a week—the cost of a house! They were shushed by neighbors who asked anxious questions about disabled family members. They were told that a program of volunteer neighbor help was to be developed so that physically disabled or otherwise incapacitated people would be assured of assistance from a neighbor in the case of an evacuation, assuming the neighbor was at home and remembered.

After a series of anxious questions about pets left behind, children in school, and the problems of the pregnant, the blind, the lame, the halt, and those with several preschoolers at home, a man arose and commented caustically on the proposed sequence of warning by bullhorn, followed by an orderly procession to the buses, which would then drive off to the safety of the convention center:

> I worked in chemical plants for twenty-two years and I'll tell you I've seen men die from chemical explosions. I was there when the walls blew down and all them men in the room except for a few blew out with them, and I was one of the lucky few . . . and it was years ago, and I'll never forget it! Let me tell you, when you can hear that whistle blow, it's too late and you've had it. (Art Tracy, 8/24/78)

His final statement was that he would stay and die on his property, or try to prevent the construction work from starting at all. His eloquent words brought a round of applause and cheers from the crowd. More than ever, the people felt drawn together. Most still assumed, however, that a combination of cooperation with the state officials and appeals to the sense of duty and justice on the part of high officials, all backed by some organized pressure from citizens, would result in their most pressing needs and fears being met adequately. They believed they had won attentive concern from the governor; they did not know that, by then, other constraints were operating, even on the occupant of that high office. It all came down to the same thing that the people worried about—the bills and who pays them.

Who Pays the Bills?

By the second week of August 1978, it was clear that individual citizens were no longer to carry the full costs for the consequences of the chemicals buried at Love Canal. Collective resources were being brought to bear on the problem. The residents were to continue to pay a steep financial and psychic toll, however, partly as a result of the way the knotty answer to the question, "Who pays the bills?" would be worked out, slowly and painfully, over the next few years. There were both traditional and new considerations about the question in this case. Traditionally, our governmental system results in fragmented fiscal and executive responsibility, as both state and federal bodies are empowered to tax and to spend and thus can control, aid, or block each other in some situations. When responsibility and mandates are not clarified, and no new dollars are committed, there are compelling incentives in many situations for officials to get the *other* governmental level or agency to provide resources. While the attendant negotiations and manipulations are going on between governmental units, the citizen, who pays taxes to all the collectors, often waits and waits.

Two other aspects were important in this new situation, however. The estimated costs for even partial, limited solutions at Love Canal began in the millions, and climbed higher by the day. Moreover, although no one was saying it aloud early in August, officials were concerned about setting precedents for responsibility in future disasters, whether or not they were similar to the Love Canal situation.

The State and the City

Although New York State may have commanded the resources to pay directly for the cleanup of the canal and the relocation of residents—however unwilling officials might be to expend the funds—the city's situation was different. They were financially strapped, and they were not empowered to raise taxes, as the state could do. Niagara Falls officials felt truly caught in the middle—faced by lawsuits, given disclaimer notices by their insurance carrier (Westmoore 11/18/78), and forced to pay a substantial portion of the remedial construction by borrowing money at high rates of interest (O'Laughlin 10/24/80). In addition, the city anticipated yearly losses of $175,000 in tax money when more than two hundred homes went off the tax rolls, and further tax-rebate decisions were pending for the people remaining in the Love Canal neighborhood (MacClennan and Shribman 8/22/78; Shribman and MacClennan 9/1/78; Shribman 10/9/78).

The city fathers wanted to protect the dual economic base of the city of Niagara Falls. Publicity about toxic chemicals could diminish the flow of tourists into the town, and such publicity could also antagonize chemical-company executives. The Hooker Chemical Corporation, for example, was considering whether to build its headquarters in Niagara Falls, and it was important to continue good relations with the company. Given their problems, the city's chief executives, led by the mayor, minimized the Love Canal problem in all public statements for the next two years, no matter how sympathetic they felt personally for the affected people.

The Federal Government

One day in August 1978, the FDAA's on-site coordinator, Norman Steinlauf, offered some rather philosophical musings when a reporter asked him about FDAA funds. Steinlauf pondered, "What is an emergency? To a certain extent the emergency is neutralized when you have moved people away from danger" (Spencer 8/19/78). The grim implications for funding a situation that might or might not be considered within emergency guidelines must have been clear to experienced city and state officials, who were already worried that an indefinite delay of the construction project while people were moving out would change the formal definition of the situation and eliminate the possibility of federal assistance for this state-mandated project.

The ideas Steinlauf expressed and the words he used were important, for he was the acting regional director for FDAA and the on-site liaison officer reporting to the White House staff. His remarks, we must assume, reflected the attitudes of those in the administrative levels above his own. In an interview a few days later, Steinlauf shared more of his thoughts. First, he stated that his agency definitely would *not* reimburse the state for the cost of purchasing homes and property. Since the only emergency was that concerning pregnant women and children, he thought that purchases of property were "totally inconsistent with the emergency order of President Carter" (Shribman and MacClennan 8/25/78). Having delivered that interpretation, he said that he would soon review the state's applications for funds for housing purchases and temporary removal of people, to see whether they fell under the FDAA statutes and regulations (Perrault 8/24/78).

As for reimbursing the city for construction costs, he reasoned, "These are both long-term solutions to the original problem and our reimbursement is figured only for short-term, immediate needs to prevent danger to life." He said that construction reimbursements would most likely have to come not from FDAA but from other federal programs, such as the EPA (Perrault 8/24/78), a view already expressed by another FDAA official (Spencer 8/19/78; *Buffalo Courier Express* 8/20/78). That cheery prospect was not congruent with an earlier newspaper headline: "Canal Crisis Shows EPA It Lacks Power, Funds." The article reported that the EPA itself needed funding to carry out its mandates, to set up regulatory programs for the hundreds of thousands of tons of hazardous waste produced annually. An official of the Office of Management and Budget (OMB) was quoted: "We can't have the President declare an emergency every time one of these bombs goes off" (McCarthy 8/11/78).

By now it was clear to everybody that the dollar costs at Love Canal were going to be high. On Wednesday, August 23, three weeks after Commissioner Whalen's dramatic announcement, the headlines read: "Love Canal Costs Put at $22 Million: Chances Slim U.S. Will Offer Much in Aid" (Shribman and MacClennan 8/23/78). State task force officials estimated the costs for the temporary relocation of 237 families and purchase of their homes at $11.6 million, the drainage system at $9.2 million, and the remainder for the health department, the DEC, the Niagara Falls School Board, and the Niagara County Health Department (MacClennan and Shribman 8/22/78).

Thus, by the third week in August—fully two weeks after the president had declared that a state of emergency existed—there was no real assurance of federal funding, even for the construction work at Love Canal, now estimated to cost many millions of dollars. As for relocation costs for even the first two rings of homes, the possibilities of federal funding, which had looked so certain on August 9, seemed to be nearly zero, although state officials still had some hopes for it (Spencer 8/19/78).

On August 9, the day funding had appeared so tantalizingly close, the announcement about the purchases of homes by the state was made while a three-hour meeting was taking place at the White House, involving members of the president's staff, high-level federal and state appointed and elected officials, and a military representative, all gathered to discuss funding responsibilities for Love Canal. Lois Gibbs, the new Homeowners Association president, had, as mentioned earlier, been telephoned the day before, and cordially invited by Willian Wilcox, head of the FDAA, to fly to Washington at government expense to join that group. For a few days afterward, she was "personally thrilled" and thought of herself as "the housewife who went to Washington." Within a short time, however, she realized and publicly stated that she had been brought in merely to witness the events and to be photographed as a citizen participant. "They made decisions, or tried to and tried not to. I was the window dressing," she commented later. "I learned how fast politicians can say one thing and then turn right around and do another." In fact, the process was somewhat more complicated and subtle than that.

He Who Pays the Piper May Change the Lyrics

With the potential problems in federal funding, state officials found themselves modifying their reasons for what had been done, to try to make their actions fundable under federal guidelines for emergency aid. The modifications in public posture might have been just another amusing example of playing the bureaucratic word game, but, in this instance, the funding constraints led to policy decisions, which led to changes in the rationale upon which policy was based, which then guided further decisions, behavior, feelings, and thoughts about the Love Canal situation on the part of the state officials who were in closest contact with the problems and with the people. The changed rationale, logical as it may have seemed for purposes of obtaining federal funding, helped to erode the trust between citizen and state at Love Canal.

At first, state officials, taking FDAA Administrator Wilcox at what they thought was his word, hoped there would be "creative interpretations" of appropriate statutes so that the home purchases would be covered by federal funds, even though this would be an expenditure virtually unprecedented in federal disaster-relief efforts. Those purchases never were funded, despite state officials' earnest efforts (MacClennan and Shribman 11/16/78; Wilcox 1/9/79), and funds proved difficult or impossible to extract from federal sources for another two years. Early in September 1978, the FDAA returned New York's first request for $22 million (see MacClennan and Shribman 8/26/78). The applications may not have been prepared properly. However, in order to understand how the rejection could happen, after what seemed to be encouraging words from both William Wilcox and President Carter, one must examine a simple syllogism of disaster funding:

1. Most disaster legislation covers natural disasters, or "acts of God."
2. The Love Canal situation is man-made, not an act of God.
3. Therefore, the Love Canal situation does not fit under most disaster legislation; therefore, it cannot be considered a disaster—and perhaps not even an emergency.

The implications for actions are three: change the rules; reinterpret existing rules so that man-made disasters are included (under a category called "other," for example); or reinterpret the current situation so that it fits the rules as they exist. In the Love Canal case, a long-term attempt at the first approach resulted in a series of federal and state legislative hearings during 1978 and 1979 for the purpose of devising statutes to assist the victims of chemical waste disasters.[2]

The second option, reinterpreting existing rules, was not employed to any extent, despite Congressman LaFalce's insistence that, properly used, legislation appropriate for aiding the Love Canal residents already existed. In a statement to a House subcommittee in the spring of 1979, William Hennessy, chairman of the state task force, addressed this point. He stated that FDAA Administrator Wilcox had termed the Love Canal situation "unique" after visiting the area on August 5, 1978, and that he "noted that it would require a somewhat creative interpretation of the disaster legislation." Hennessy also pointed out that, when President Carter declared the emergency, he "authorized such Federal actions as were deemed necessary to save lives and to protect property, public health and safety or to avert or lessen the threat of a disaster." At the White House meeting on August 9, 1978, Hennessy went on, state officials had informed federal officials of the magnitude of the problem and had claimed that permanent relocation of some 250 families would be necessary. "*We received reassurance that there would be*

Federal participation in the temporary relocation, that consideration would be given to Federal participation in the cost of permanent relocation and that Federal assistance would be provided for the remedial construction." (USHR 3/21/79, pp. 153-154, emphasis added). The mix of federal, state, and local funds anticipated was described by Commissioner Hennessy (see table 4-2). The expenditures, which became greater over the next year, will be listed later.

Why were federal officials reluctant to reinterpret existing rules? Some would argue that the federal government, particularly the Office of Management and Budget, was simply reluctant to engage in setting new precedents when the potential costs and implications were unknown. Some interpretations required imaginative stretching of rules, an exercise some critics say the "old-line disaster establishment" simply will not aspire to. If the Love Canal situation were truly an *emergency*, traditional reasoning went, then essential cleanup and other remedial measures for the site and people should be accomplished within sixty days. If they are not done within that time, then, by definition, an emergency situation never existed. The Love Canal remedial work was not begun until some seventy days after the health commissioner's order—proof, under this reasoning, that the Love Canal situation, which had taken some thirty years to develop, simply did not qualify under the existing definition of an emergency.

Another issue was involved—long-term federal and state relationships and responsibilities. William Wilcox reiterated on-site FDAA coordinator

Table 4-2
State, Local, and Federal Funding Commitments to Love Canal
(thousands of dollars)

Purpose	*Total Cost*	*State of New York*	*City of Niagara Falls*	*Federal Government*
Remedial construction—southern zone	5,000		3,600	1,400 (FDAA)
Remedial construction-northern and central zones	4,650			
Health and environmental testing and services	2,725	4,258		4,000
Temporary relocation	883			
Other	800	200		600 (FDAA)
Permanent relocation, including acquisition of homes in rings 1 and 2	9,216	9,216		
Bus evacuation plan during construction	550	550		
Human-services grant	200	200		
State aid for property-tax relief	1,000	1,000		
Total	25,024	15,424	3,600	6,000

Source: Statement of William C. Hennessy, commissioner of transportation and chairman of the Governor's Task Force on Love Canal, State of New York (USHR 3/21/79, p. 156).

Steinlauf's narrow interpretation of the mandate provided by disaster legislation, saying that, if the FDAA were to provide assistance to the state for the Love Canal situation, assistance might have to be provided for a period of years. Not only was this factor not contemplated in the FDAA legislation but, furthermore, "the effect would be to discourage the appropriate and responsible local, state and federal agencies from developing a coordinated, consistent and long-range solution to the Love Canal problem and such similar problems as may exist at other locations, either nearby or at some distance" (USHR 3/21/79, pp. 153-154). To the people of Love Canal, his reasoning was similar to that of a lifeguard who might refuse to pick up a struggling, drowning person because the act might impede the development of a good water-safety program in that person's community at some time in the future.

The third option—reinterpreting the Love Canal situation so that it fit under existing rules—was finally resorted to by state officials. Home purchases for health reasons were definitely ruled out of bounds by existing disaster legislation, but there was still the possibility of obtaining funds for purposes related to the construction work. By the time the Love Canal task force resubmitted its application for FDAA funds in October 1978, they argued: "The state position is that FDAA participation in the cost of purchasing the affected homes together with permanent relocation costs incidental to closing is within the intent of the law and the emergency declaration *because these costs were incurred as a direct result of remedial construction*" (emphasis added). The paragraph went on to explain that trenching had to be done in the backyards of houses in ring one (NYDOT 11/78).

The sensible-sounding idea clearly conveyed by Health Commissioner Whalen's August 2 order and by authoritative reassurances offered during August—that, when people moved, it was in order to protect their health—was soon officially a thing of the past. It seems that he who pays the piper not only can call the tune but can change the lyrics as well. Those changed lyrics became part of the well-orchestrated song played ever more loudly after the primary election was safely behind the incumbent Governor.

Since federal funds, if available at all, would pay only for the remedial construction project and directly associated costs, the state officials emphasized more and more that homes had been purchased not because of the health hazard but to allow the remedial construction to proceed, with pipes and trenches to be placed in the backyards of the homes that abutted the canal. This new rationale did not sit well with people who remembered well Commissioner Whalen's order depicting the canal as "an imminent peril to the health of people," who relied on Governor Carey's promise to consider moving more people if evidence of further health hazards were forthcoming, and who were quick to point out that the new rationale did not explain the purchase of the ring-two homes whose yards did *not* abut the canal.

Under FDAA rulings, in order to obtain full federal reimbursement for the construction project, which began at the southern portion of the canal, the work had to be completed within six months of the president's declaration that a state of emergency existed. The president made that declaration on August 7. Two months were occupied in developing contingency evacuation plans and other safety plans. Consequently, once the remedial work did begin, there was great pressure to complete the first phase in four months, before the February cut-off date for funds. Crews arrived before dawn and worked full-tilt, often up to twelve hours a day, seven days a week, carrying out procedures that were not usually done in winter weather and regardless of the original plans to confine the work time to the hours when the adults in the community were likely to be at work and school-age children were out of the houses (see Clement Associates 8/20/80, p. 3).

The paradoxical notion that a multimillion-dollar construction project had been initiated to cover a leaking chemical-waste dump site, that 237 families were moving out and selling homes at government expense, that costly health and environmental studies were commencing, but that these matters were somehow not related to concerns about health, became a part of the explanations shared by the government officials. The explanation seemed to turn into a belief about the true nature of the Love Canal problems. Such a belief would be consistent with the state's continued use of the 99th Street School as headquarters, drawing hundreds of people to the central portions of the canal site for more than seven months, the easy public access to the construction site during that period, lax safety precautions on the part of workers, and a general attitude on the part of government officials of minimizing the possibility that the chemicals buried at Love Canal might actually harm anyone physically.

The Two Cultures

During the late spring and early summer of 1978, the formal meetings between residents and governmental officials had provided the public context within which the residents began to feel part of a cohesive group, encountering problems they thought were caused, in part, by the very officials and agencies who saw themselves as trying to assess and offer help for the problems caused by buried chemicals. From early August on, the interactions between government officials and Love Canal residents were, with increasing frequency, in the form of meetings and negotiations between representatives of organized entities, rather than the individual, somewhat haphazard encounters that had characterized government-resident relations in the early spring. As both entities became more organized, statements made during any formal meetings, remarks passed in unplanned daily encounters, descriptions of ongoing events, printed and oral reports, information about the personalities of various key people, knowledge about the problems residents experienced with their health and finances, attempts to get resources, problems the state was having in finding outside funds, all became woven into the cultures of what was rapidly becoming two distinct, recognizable, opposing sides. Ideas about appropriate goals based on what the problems *really* were at Love Canal and how to go about solving them began to differ markedly when viewed from the different vantage points of the resident and the governmental official.

There were differences within each group, depending again on individual circumstances, but the differences between the two groups far outweighed any internal tensions. Above all, most residents wanted to leave open the definition of how much harm had been done to their environment and to their health. They wanted the decisions about essential help left flexible, so that appropriate help could be offered to fulfill needs as they became apparent. The officials wanted the definition of harm to be narrow, to fit the resources they had available, so that some tasks could be successfully accomplished, accounted for, and pointed to. In a situation with so many unknowns, the people wanted to feel that the available help would be sufficient to meet their needs, even if some of their needs could not yet be anticipated; the officials wanted to define some aspects as *known* and to address their attentions to these. Just as they were to fence in a defined area around a construction site when the underlying chemical leachates might be flowing far beyond, so they wanted to define and bound the problem to be solved, even though it lay within the problem that in fact *existed.*

As time went on, the separate clusters of shared ideas gave different meanings to words and to the events as they occurred. The gap between the two sides grew, slowly at some times, more rapidly at others. Most people, however, in those first months of August, September, and early October 1978, continued to trust that the governor and the health commissioner were above the daily fray of the sometimes abrasive relations between the harried on-site interagency task force workers and the nervous, frustrated residents. This feeling of trust lasted until the third week in October, when an episode occurred that opened the eyes of many of the residents to a new set of realities.

The Great Awakening

The safety plan was never formally approved by the residents, but the construction work did proceed. Although the plan was not formally approved, neither was it formally disapproved, nor did the largest organized group, the Homeowners Association, support an attempt to prevent the start of the construction work. That attempt was made by a group of tenants from the nearby rental units, who were dissatisfied with the state's lack of attention to their special needs. They sought some leverage by requesting the court to order an injunction postponing the starting date for the construction project (Shribman 10/8/78). The move was not supported by the Homeowners Association officers, who reluctantly agreed with the state that something had to be done to stop the chemical leaching before the cold winter season. The association leaders bargained with the state task force representatives and chairman, winning a promise that criticisms about the drainage plan and assessment of leachate migration would be attended to—a promise that could be postponed. They were promised—and that promise was kept—that the state would hire a consultant for the residents, a toxicologist-chemist of their choice to report to the residents as their independent consultant, even though the state would pay his salary. (Stephen Lester, the consultant who arrived on October 10, proved to be a valuable resource person for the residents.)

One most important promise clinched the association officers' decision to acquiesce in the work start-up date. That promise was implicit in the governor's statements that decisions were still pending about moving families out of the area temporarily if their health might be affected by fumes released during the course of the construction (Shribman 9/26/78).

In the weeks before the construction finally began, people from several dozen families came to the Homeowners Association headquarters, phoned association president Lois Gibbs, and spoke with each other and any interagency task force staff they could contact about those pending decisions. Gibbs had been told that the health department's evaluations and recommendations on the first cases would be available on October 11, the day after the construction work was to begin. The first cases were those of nineteen families in outer-ring homes, who had submitted their complete medical records and requests to be moved during the construction period, on the grounds that the release of fumes might exacerbate their health problems.

On October 10, crews of workmen with machinery arrived, and a ceremonial shovelful of dirt was turned at the construction site (Shribman 10/10/78). The construction work had officially begun.

On the next day, an anxious Lois Gibbs telephoned Health Commissioner Whalen from the association office to ask about the health department evaluations and recommendations on the first nineteen families. She wanted to arrange to be with some of the people whom she knew to be most upset, in case any of the requests were turned down. She knew the concerns of the people well—from their repeated inquiries and long conversations with her at all hours of the day and night. Three of the families were already living in motels, through an arrangement between the interagency task force chairman and the United Way of Niagara Falls, an acknowledgment, in her opinion and that of the families, that there indeed were health hazards at Love Canal for certain families living beyond the inner ring.

When Gibbs reached Health Commissioner Whalen with her request, he told her that telegrams would be sent to the families to explain what the decision was in each case. Such confidential information, he chided her, could not be imparted to anyone but the families. Gibbs returned the telephone receiver to its cradle, where it was picked up immediately by news reporter David Shribman, who had been standing next to her. He dialed Commissioner Whalen, who promptly told him that the decision had been made that relocation was not warranted for any of the people (Cerrillo and Gibbs 10/10/78; MacClennan and Shribman 10/12/78).

The next morning, October 12, the nineteen waiting families received identical telegrams:

> Dear ______________:
>
> All of your medical records submitted to my office have been thoroughly examined. Our review of this information and all currently available data relating to the Love Canal, the planned remedial construction and associated evacuation and safety programs has resulted in the decision that temporary relocation is not warranted at present. If you have any additional information you wish to present, please do not hesitate to write to me. Meanwhile, we will continue to monitor and evaluate the progress of remedial work at the Love Canal site for potential hazards.
>
> Sincerely,
> Nicholas Vianna, M.D., M.S.P.H.
> Director
> Occupational Health and Chronic Disease Research
> N.Y.S. Department of Health
> (NYDOH 1979, Sect 58)

The residents were shocked and then furious at what they dubbed "diagnosis due to address." They rushed in a body to the health department office, demanding that the physician in residence that week explain the telegrams, which did not address the specifics in each case. It was not a pleasant scene in the health department office, where a large crowd of enraged, frightened people confronted a physician who appeared equally frightened. Dr. Phillip Gioia, new on the site, could only let them telephone Dr. Glenn Haughie, the health department's deputy director, in Albany. An old hand, Haughie calmly told the callers that the decisions were final but that, if *new* information were submitted, the cases would be reconsidered.

This episode marked a significant point in the people's growing mistrust of the intentions of the state. At first, the people, long accustomed to what they viewed as indifference or worse from the city and county officials they encountered, had wanted to trust all the state experts who had become visible on the scene in the spring of 1978, seemingly ready and eager to help. The people's trust faded step by step; first they began to mistrust the lower-level workers, then the elected politicians and high-level appointed officials, and finally, as we shall see, the scientists and physicians who were involved in the situation on an official basis.

Notes

1. The other task force agencies were the departments of banking and insurance, and the divisions of housing and community renewal, equalization, and assessment, all involved in matters related to the purchase of homes. The office of disaster preparedness was deeply involved in the preparation of a safety plan related to the construction project, and, because federal funds were supposedly available for emergencies, they were involved in some negotiations with the Federal Disaster Assistance Administration (NYDOH 9/78).

2. As a consequence of political compromises, the legislation to assure the cleanup of hazardous wastes (P.L. 96-150), popularly called the Superfund Bill) finally passed both houses of Congress with *no* provisions for compensating human victims of such wastes. Love Canal was referred to frequently in congressional discussion, Love Canal residents and involved officials and consultants testified at legislative hearings, and Senator Moynihan publicly lauded the organized citizens of Love Canal (NBC-TV, 11 PM News, 11/24/80). He commended them for their part in bringing the problem of toxic wastes to the national attention and for keeping it there. Senator George J. Mitchell of Maine said, in criticizing the bill, "By what standard of justice or decency is damage to property more important than damage to persons?" (Lynch 11/25/80).

5 The Health Department and the People

The New York Department of Health (DOH) played a major role in identifying Love Canal as a health problem of potentially enormous magnitude. Dr. David Axelrod deserves credit for sounding the alarm about the health hazards at Love Canal when he was director of the DOH laboratories, before he became New York's health commissioner. Commissioner Robert Whalen's orders set in motion the events that made Love Canal the symbol for the widespread toxic-chemical-waste problems in our society. The relationship between the DOH and the Love Canal residents, however, soon turned into a bitter struggle.

On October 12, 1978, two days after the massive remedial construction project formally began at Love Canal, the Love Canal community shared the shock of the nineteen families whose health records had been examined by the DOH and a committee of consultants and who had received form telegrams from the DOH stating that temporary relocation was not warranted for them at that time. Many residents now felt betrayed by the department of health. They were suspicious that the decision that relocation was not warranted was based only on financial and political realities. Now they felt that previous assurances that their health problems would be considered individually had been offered in order to dampen their public resistance to the already delayed commencement of the huge construction project, which in turn could have led to the loss of federal funds promised for that work.

The DOH professionals, however, may not have considered themselves involved in policy decisions about Love Canal people, but rather as researchers whose major task was to add to the store of knowledge about the action of chemicals on living beings. Although they might strive to produce assessments of the increased risks faced by various-sized populations of contracting chemical-related disorders, even risk assessment could be viewed as an extension of their basic research for facts, with policy decisions to be made by others, who would use scientific information in making such decisions if they so chose.

The DOH had launched its extensive studies in response to the health commissioner's August 2, 1978, order to undertake epidemiological studies beyond the preliminary ones done in the spring and summer of 1978, in order to "delineate chronic diseases afflicting all residents who lived adjacent to the Love Canal landfill site" (Whalen 8/2/78*b*). The order said nothing about how such studies would enter into policy considerations—simply stating that studies were to be done. A report was to be submitted to the governor and the legislature by September 15, 1981, in compliance with a legislative act that allocated $500,000 for such studies (N.Y. Public Health Law, Sect. 1389, 6/22/78).

Most of the Love Canal residents assumed that the research data would be the basis for reliable information to be relayed to them about the condition of their health after their long exposure to buried chemicals. Only with that kind of information could they plan to alleviate or live with their problems. Although they also expressed the understanding that "scientists studying here can learn things to help other people sometime," they felt an immediate need for expert help. The DOH scientists and physicians, however, seemed to think that their job was to gather the data and analyze them. They did not seem to view themselves as having clinical responsibilities for the affected population.

The differences in the perceived goals for the DOH work at Love Canal, the uncertainties about the impact of the chemicals, the social and political context of the work—where decisions *were* being made and would continue to be made about the fate of the Love Canal residents—the DOH's conduct of the early research studies, the kinds and timing of the DOH official announcements, all contributed to the development of a deep animosity between the DOH leaders and the Love Canal residents.

Uncertainties and the Residents' Need for Information

New York's Department of Health prides itself on being one of the very best in the nation. It is well staffed and equipped to handle most of its mandated tasks; in 1978 it had a $124 million budget, 6,000 employees, and extensive laboratory facilities. (*New York Times* 12/17/78). The department's physicians, other health professionals, and research scientists consider themselves members of an organization akin to a university or a research institute.

The opportunity to do research at Love Canal presented quite a challenge to a health department that prides itself on scientific research capabilities. Love Canal was the first large residential area known to be affected by such a huge quantity of buried chemicals, a good number of which were known to have serious toxic effects on the human body. Knowledge about the effects of most chemical compounds on the human body is relatively meager. Hundreds of chemical compounds have been studied, and their effects on animals have been observed, but the number of compounds already studied for effects on animals or people, is tiny relative to the numbers of chemical compounds in use and constantly produced by the chemical industries (Council on Environmental Quality 1980). For New

York's Department of Health, a great research opportunity was now at hand, accompanied by great problems.

Usually, under carefully controlled laboratory conditions, animals are exposed to chemical compounds one at a time, and extrapolations of the results in the animals provide estimates of their effects on human beings. In contrast, at Love Canal, by the summer of 1978 eighty chemicals and by late autumn more than two hundred had been identified in the leachates coming from the canal site. There were and still are insurmountable problems in estimating precisely what exposure people had to chemical substances. Various compounds had been disposed of in different locales and had migrated to new places in the canal and the neighborhood. The compounds may well have undergone changes, through interaction, synthesis of new compounds, and possible neutralization or activation of others. With our present state of knowledge, the potential effects of compounds produced by such interactions are unfathomable.

When the New York Department of Health undertook its studies at Love Canal, some DOH scientists first tried to compare the results to Occupational Safety and Health Administration (OSHA) standards. This proved fruitless, because OSHA standards for acceptable conditions in the workplace are designed for people of working age, usually male, healthy enough to be employed, and exposed for up to forty hours a week to chemicals whose dosage and nature are more or less known. Love Canal residents, however, were people of both sexes and all ages, newborn to elderly, in all conditions of health, and possibly exposed night and day, for years, to unknown amounts and combinations of chemicals.

The DOH did arrive at some important conclusions about health hazards in the summer of 1978 and reported them to the governor, the state legislature, and the general public. The Love Canal residents assumed that, at the state capital in Albany, the commissioner of health and other high-level scientists would continue to confer with the governor about ways to protect them from possibly serious health effects in the long run.

On a more immediate basis, in Niagara Falls, the people living in a place that had been declared a hazard to their well-being felt entitled to help from their department of health. Although the residents agreed with the DOH view that the department was not to replace the residents' family physicians, the people did see the DOH physicians and researchers as possible sources of information and advice that was unobtainable from other sources. For a time they expected to have a working relationship with the DOH professionals that would be superior to their usual physician-patient exchanges, for the residents perceived that the DOH had exercised greater wisdom than that of local doctors when they recognized that the old canal was dangerous to the residents' health.

The DOH became formally involved at Love Canal in the spring of

1978. Because the people wanted to know what had happened to them, what might happen to their children, and what was going to happen to their homes and neighborhood, they were eager to cooperate in any way possible with long-term testing programs and were ready to offer the DOH researchers their own experience-based information. At first, openly sympathetic state health officials imparted information to residents both publicly and privately. However, the DOH's early handling of both the first round of health studies and the early informational meetings resulted in fear, confusion, and antagonism on the part of the residents. As time went on, communications dried up in the heat of angry public reactions, often shouted at officials in expressive, earthy terms. State officials on site began to guard their words, either of their own volition or as part of their agency's policies.

The emotional public meetings were later cited by the task force coordinator and others as the major reasons for the emotional and physical distance that the DOH professionals put between themselves and the Love Canal residents. There was a more basic reason, however, a prior attitude of the DOH professionals that was at variance with the residents' assumptions.

The DOH evidently did not consider providing help by giving solid health information or helping people with their fears about their health as important and appropriate purposes for the department's work at Love Canal. Because the DOH officials did not pay serious attention to the task of providing information to them and working through the implications of the information, the residents felt that they were being treated not as rational, respected adults but rather as though they had somehow lost their mature good sense when they became victims of a disaster they had had no way of preventing. (See Averill 1979 for a discussion of the importance of information in reducing stress.)

When, by chance, an official was on hand and willing to talk with people and patiently to repeat explanations about bewildering new technical matters, the people's fears were temporarily mitigated. For example, Dr. LaVerne Campbell helped establish the health department's on-site office after the commissioner's order of August 2, 1978. He was remembered by the residents as a sympathetic person who made himself fully available during that period to explain technical matters to everyone. However, a few weeks after the DOH office was established at the 99th Street School with all the other agencies' offices, few high-level DOH specialists were at hand to supply information to the people. After Dr. Campbell's August and early September tour of duty, state physicians took assignments at Love Canal for short periods, often for only a week. Even if they were sympathetic and concerned, new physicians were not familiar with details of the rapidly changing situation or with most of the residents. By early September, homeowners and tenants were complaining to each other, to the Homeowners Association officers and volunteer office workers, to the state task force on-site coordinator, and to anyone else who was interested, that they did not understand technical details about the health and environmental studies or the forms they were to fill out and that they were bewildered about the test results they had already received. They complained that the state doctors either did not seem to know much about the situation or did not explain things to them satisfactorily.

The DOH people, however, seemed to be there not to offer care or information, nor to work with people personally, but rather to collect data for long-term epidemiological studies, according to a predetermined research design.

The Residents Seek Outside Experts

As early as mid-August 1978, the organized residents asked the president of the large state university nearby in Buffalo to locate faculty members who could offer them information and advice. They wanted the help of experts who were not part of the department of health or the rest of the governor's interagency task force. The reasons for the university's administrative response will be discussed later, but it was minimal and useless to the residents, who had immediate needs and concerns. Some very important help did come from a few outside experts from academia, however, who, as individuals, volunteered their time and efforts. In the summer of 1978, Dr. Wayne Hadley, Lois Gibbs's brother-in-law, not only helped Gibbs to assume a leadership role in the Love Canal community but also alerted two other scientists, Dr. Charles Ebert and Dr. Beverly Paigen, to the Love Canal situation. Like Hadley, both Ebert (a geographer and expert on soils science) and Paigen (a biologist, geneticist, and cancer researcher) were keenly interested in environmental issues, were concerned about this particular disaster, and were willing, among other things, to make technical terms understandable to lay people. Hadley moved to the far west in early September 1978, and his connection with the situation became attenuated. Ebert's advice was confined to matters concerning the construction project. Only Dr. Paigen was involved in health research, and she became vital to the Love Canal residents and was an important factor in the entire story, as we shall discuss in detail in the next chapter.

Paigen's first encounter with the residents' clamor for information occurred at a meeting she attended, at Hadley's invitation, on July 19, 1978. Here, health department officials distributed slips of paper to homeowners, listing the polysyllabic names of ten chemical compounds. Many names had numbers next to them representing the values for air readings taken in the residents' homes (table 5-1).[1] The officials conducting the meeting frustrated and baffled the audience when they could not provide answers to

Table 5-1
Air-Reading Values in Four Love Canal Homes
(micrograms per cubic meter)

Compounds	*House #1*	*House #2*	*House #3*	*House #4*
Chloroform	4	99	10	
Benzene	188		107	11
Trichloroethene	24		44	4
Toluene	5,703		1,160	45
Tetrachloroethene	62	1,472	6	4
Chlorobenzene	154		115	
Chlorotoluene	6,700		340	
m-p-xylene				
o-xylene				
Trichlorobenzene				
Total	12,835	1,571	1,782	64

such prime questions about the meanings of the figures as "Is it safe to be in my home?"

Paigen asked the woman seated next to her if she could see the list the woman was staring at, and then she pronounced the chemical names when the woman asked her to. "Are you a scientist?", the woman whispered. Paigen nodded in reply. "With them?", she gestured toward the stage. Paigen shook her head, and whispered "No." By the end of the meeting, word had spread through the crowd that an outside scientist was present. The Love Canal residents surrounded Paigen, begging her for an interpretation of the raw data they had been given. Although she was not able to provide an interpretation of the numbers, Paigen remained after the meeting, taking time to talk with people and to explain what the various chemical names meant, how and why sampling procedures were done, and the difficulties of interpreting the results.

The lack of information or, perhaps more subtly, the lack of recognition on the part of the health department that the people felt entitled to useful information from knowledgeable authorities, was a constant irritation for many residents. Beyond the department of health's basic stance on their own work priorities, it seemed that the general ethos of the state's efforts was to mitigate the Love Canal situation almost solely by imposing technological and other practical solutions on the area and the people, without considering the people's reactions. A discussion of the state task force, the administrative structure created especially to handle the Love Canal problems, will help us understand the attitudinal context within which the DOH worked at the Love Canal.

Getting the Job Done

On August 4, 1978, the governor's Love Canal Interagency Task Force became an official entity. From most reports, on-site personnel from the lead agency, the department of transportation, handled the delicate issues of negotiating the Love Canal home purchases in the fall of 1978 fairly and sensitively. In general, the residents said that the whole task force, under the guidance of Mike Cuddy, the on-site coordinator, "did the job they had decided to do."

The interagency task force originally was intended to be a flexible administrative agency. However, once goals and tasks were decided upon and work routines established, production took precedence over flexibility. By mid-August, the task force focused on three goals: (1) fulfilling the governor's promise to purchase homes located in a circumscribed area and to help their owners move first to temporary housing and then to new homes; (2) undertaking studies to assess the damage to health and environment; and (3) supervising a massive construction project. Given the perceived financial constraints and opportunities, these three emerged as feasible goals during the early weeks of August, then hardened quickly to become the chief or only goals of the members of the Love Canal Interagency Task Force (Cornwell 1980). There seemed little, if any, intent to anticipate and alleviate major problems that might become apparent for those Love Canal residents who remained in the neighborhood after the August decisions were made and announced.

Just fulfilling those goals resulted in an enormous workload for the task force. During the first few months, long working hours, including evenings and weekends, were routine for the task force. The department of transportation's real-estate division, for example, accomplished a normal year's work in a few months. Problems and snags arose every day for the task force. Some problems were inherent in the tasks to be accomplished—including problems with landlords, worries about funds, problems with insurance companies, and a miscellany of concerns of varying degrees of seriousness, all of importance to the residents who were suffering through these difficult times.

Locating and moving to temporary quarters provided special problems for the residents and for the task force. Some landlords were uneasy about renting to people who would be leaving within six months and who offered rent vouchers from the state, not cash, as their means of payment (Komorowski 9/26/78). Although rental units were available at a nearby Air Force base, those who moved into them found themselves under unaccustomed military restrictions. One pregnant woman called the DOT one day, weeping. She complained that her mother, who was to stay with her to help care for her other children during and after the birth of a new baby, would not be able to park her car near their apartment; parking spaces were assigned to residents, and parking was not allowed on the streets. The apartments had to be cleaned to military specifications before people left. In one instance, when a family wanted to stay longer than originally contracted, they were told they would have to clean the quarters thoroughly, move out, and then move back in. This may have been a misunderstanding on this family's part, but that is what they reported and what a task force person had to straighten out.

There also was always the hint of violence. More than once, angry residents descended on task force headquarters in a body. Although no acts that could be labeled personal violence actually occurred, the threat was always there. There were also the ordeals of public meetings. The threat of publicity for every blunder was a real one, and every word spoken by task force officials had to be said guardedly. A delicate racial problem might have been stirred up easily, too, because renters, many of whom were black, believed that the white homeowners were getting the lion's share of attention from the state, and homeowners, almost all of whom were white, felt that renters could and should simply move; and both sides spoke their minds freely.

A variety of other problems arose in the task force's everyday work. When banks were reluctant to grant mortgages to older people; when a bank wanted fifty-dollar deposits for access to their title searches; when insurance companies wanted to cancel homeowners' insurance thirty days after they had moved out but while they still owned the homes; when school personnel refused to allow the children to eat lunch in school, in order to keep them out of the neighborhood during construction hours; when records were lost or misplaced; when thousands of gallons of leachate suddenly poured out into the street (Komorowski 3/10/79); when the state accepted the offer of a local institution to allow handicapped and retarded clients to work, at minimum wages, clearing overgrown grounds near the construction site, with little provision for their safety (*Buffalo Evening News* 5/11/79)—then the task force personnel in Niagara Falls, or at least Mike Cuddy, the coordinator, and his staff, heard all about it, sometimes very forcefully.

The task force members not only had to solve problems but they had to develop thick hides as well. Every day they worked to troubleshoot a multitude of daily problems while maintaining a harmonious appearance among members of nine different agencies and carrying out policy decisions made by officials who were hundreds of miles away at the state capital. Recognizing the importance of that higher level, Gibbs kept in direct touch with the authorities in Albany, often bypassing the local personnel completely. She justified what might have been taken as an affront by the on-site task force coordinator by saying, frankly and frequently, "The people on-site can't make decisions, so why waste time? I'll talk to the people with some power."

Although the on-site task force responded flexibly to problems that arose as they tried to accomplish the three goals that became clear by mid-August, major policy changes after August were concessions wrested from reluctant officials after enormous efforts by residents, who used every tool of persuasion they could muster. The changes were certainly not part of a state-mandated plan to respond flexibly to new needs as they were discovered.

When the various agencies were assigned to the task force, there apparently was scant anticipation that there might be deep emotional traumas among people who were suddenly selling their homes for fear of their health and safety and among other people who were left behind. When one cost estimate for Love Canal was published in mid-August, the $34,000 allotted for human services from a budget of $22 million, represented $15.50 out of every $10,000 of planned expenditures (Shribman and MacClennan 8/23/78). When the social service department was invited to join the Task Force, it was not to alleviate the social and emotional problems of residents but to respond to the needs of the DOT personnel, who called for help when they found themselves swamped at the Love Canal scene (Keys 12/15/78). The social service department's trained, experienced interviewers were called on to assist in the relocation of more than two hundred families. Upon undertaking the interviews, the DOT-social services teams used standard DOT forms, adapted from normal DOT operations (that is, home purchases in relation to highway construction). There were no provisions for information about the residents' personal problems or special financial, health, or other needs, other than those routinely encountered in DOT transactions (Keys 12/15/78). The lack not only led to some increased work and confusion for the social service workers and some residents but also showed that the very special circumstances of Love Canal had not been fully taken into account.

The social service workers who joined the effort were not counselors. The first ones came from an auditing and quality-control division within the social service department. With their major task at Love Canal to help relocate families, it was their ability to work with community facilities and governmental agencies that was valued. Once on board, they joined with the others, working heroically as part of a crew trying to get a huge task well launched. That the social services workers did yeoman duty, working nights and weekends for months, is an indication of the need they filled, but attention to social and emotional factors was not a core part of their assignment. Little if any thought was evidently given to serving the residents better by anticipating an emotional drain on overburdened, harassed, on-site task force personnel. Students of service-providing organizations know that personnel who work routinely with great numbers of stricken people can easily turn into "heartless bureaucrats." They develop the viewpoint that the people they are helping are units to process if they are to get their job done. The development of an *esprit de corps* does not seem to help in these cases, but in fact may serve to separate the workers even more from the feelings of suffering people.

Such a processing mentality developed at Love Canal, despite the widespread, genuinely sympathetic feelings that task force workers had for the Love Canal residents. For example, one day an ordinarily patient task force worker forcefully expressed his irritation with a "picky, spoiled, stubborn" young woman, for whom he had located a temporary apartment. She was undecided about moving into it and told him she might move into it but then move to another location shortly thereafter. He told her firmly that the state would pay for one and only one move to a temporary place, barely concealing his irritation as he spoke to her. The young woman was pregnant and desperate to flee the contamination she feared had already injured her unborn child. Since her parents lived several states away, she wanted to move immediately to protect her baby and then move again to locate near other relatives if a suitable place became vacant. Her problem in making a choice was viewed by the worker as a willful, unnecessary obstacle to his efficient performance of a job that he was doing under very tough circumstances. Feelings such as his became less uncommon as time went on, and the sympathetic, well-intentioned task force workers increasingly viewed the residents—particularly the articulate, organized residents—as "necessary nuisances" who made it more difficult for the task force to achieve its set of well-defined goals.

The DOH was embedded in this structure and, like the other agencies, had a clear goal. The DOH goal simply was not the one that residents assumed a department of health would have in this situation. The DOH scientists wanted to study the health effects of chemicals; the people wanted those studies to be *used* to help them, immediately if possible, were they found to be harboring the effects of chemicals.

There was yet another pervasive reason for the strained relations that developed between a great many Love Canal residents and the DOH professionals. At the time of their very first experiences with the DOH research efforts in the late spring and early summer of 1978, residents were patient with the DOH's apparently uncoordinated, awkward moves as that public agency undertook the major new efforts entailed by Love Canal. After a brief time, however, residents came to doubt the ability of the DOH personnel to perform the core task they had chosen to pursue—the epidemiologic survey and laboratory studies of the health of the Love Canal people. As time went on, the residents became concerned that their futures were in the hands of people whom they viewed not only as cold and distant but actually as incompetent at the job they had set out to do. Whether valid or not, this perception added immeasurably to the load of worries carried by the people of Love Canal.

The Health Department Starts the Health Studies

Despite the great resources of the New York State Department of Health, it had no special unit for environmental emergencies. For that reason, when the health commissioner's August 2, 1978, order mandated long-term studies, the bureau of occupational health and safety undertook the task, expanding on procedures that were already in place for data collection. As the chief epidemiological investigator for the Love Canal work, Dr. Nicholas Vianna had about ten days in the spring of 1978 to prepare the first Love Canal studies, which began in earnest in June 1978. He and his staff decided to collect information about Love Canal residents' health by means of questionnaires, confirmed by physician and hospital records, and a blood-testing program. They collected the information only on Love Canal residents, with no information on groups from other regions for comparisons after the data were all in—a serious flaw in the research design from the beginning.

Blood Testing

The blood-sampling program started in June with inner-ring residents only. When other people learned of the sampling and became alarmed, a general invitation was issued to any Love Canal residents who wanted their blood tested to come to the 99th Street School. The people came, by the hundreds, to pass some of the hottest days of the summer packed in the school auditorium or in the crowded corridors. The residents' opinion of the experience was summarized in testimony before a congressional committee, some nine months after the blood-testing program took place:

> The blood testing program was never organized to any degree, it was inefficient, and extremely trying for the residents. . . . Two to four technicians were available to draw the blood samples. Hundres of people were lined up daily for testing. Consequently, this caused more stress among residents, standing in lines for undetermined amounts of time (possibly being turned down once reaching the front of the line because of shortages of needles, etc.). The State could have made this procedure better had they used a little thought and organization (plans). They could have taken people by streets or alphabetically. (USHR 3/21/79, Gibbs testimony, p. 91).

Once begun, the massive blood-testing program (4,000 samples within a few weeks) overloaded the state's laboratory facilities. (Kistler 11/9/78). Residents waited for weeks for the results of the tests. In many cases, they learned that their blood had to be retaken and retested to check the accuracy of results obtained from samples that had been kept too long or improperly stored before laboratory personnel were able to perform the analyses. Although the residents were later reimbursed from a special fund set up by the state, initially they had to pay for the retests themselves if they were done by private physicians. The results of the DOH testing were sent to the residents' physicians, adding to the residents' problems, although the reasons for this decision seemed sensible to the DOH. The massive blood sampling thus caused the residents discomfort, resulted in the inefficient use of the state's facilities, and yielded poor service to its citizens. The delays and confusion about blood tests inspired inquiries at public meetings and dozens of telephone calls to the DOH and to the Homeowners Association headquarters. In the low-income housing project, the blood-testing program was late in starting and reached only a small percentage of the residents. In sum, people were already worried, and the blood-sampling program sparked their mistrust of the very department they thought would make or at least influence important decisions about their futures.

Surveying

The people's responses to the questionnaires were poor at the beginning. Some basic care early in the distribution and construction of the questionnaires might have revealed some reasons for the poor response rate. The questionnaires were distributed in several major stages—the first ones to inner-ring families starting in June, when their blood samples were taken. The questionnaires were given out on one day and picked up the next by health department workers, who looked at the questionnaires and urged the respondents to fill in any blank spaces. Although most ring-one families were reached ultimately, there was some confusion in the process. One woman reported, for example, that, when a family friend dropped by one morning to return a borrowed power tool, he was handed a questionnaire as he walked out the front door on the assumption that he lived in that house.

The next wave of questionnaires was handed out even less carefully. The forms were simply left at the 99th Street School for people to pick up as they wished when they had their blood samples taken. In one week in August, 1,300 questionnaires were distributed in that manner (Hiltzik 8/12/78). Some people completed the questionnaires on the spot as they waited in line, with children, or in the crowded auditorium. Some used their friends' backs as desks. Early in September, social service personnel tried unsuccessfully to locate college student volunteers to distribute questionnaires to the low-income project where many minority and elderly people lived. As with the blood sampling in this geographic area, the data yield remained low, despite complaints from the local leaders and some extra efforts by the task force.

The third set of questionnaires was distributed door to door in the area east of the canal, when the survey was extended beyond the inner rings (after the August 2 order). If someone was home when the field worker knocked, the worker gave the resident a brief explanation of the study. If no one was there, questionnaires were left behind the storm door or tucked into the mailbox.

In early September, Dr. Vianna asked Lois Gibbs if the Homeowners Association would cooperate with him, because the response to the questionnaires and to other requests was so poor. The Homeowners Association board of directors decided immediately that this was a most worthwhile activity. Eventually, the volunteer office staff reached most of the residents in the area and urged them to cooperate by filling out the questionnaires, signing release forms, and asking their physicians to forward medical records. The association workers explained that the study was important, and they responded as well as they could to their neighbors' many questions.

The first question often was "What questionnaire?" because there were no instructions on the form. There was little indication of the purpose or the source of the instrument, other than the statement labeled "authorization to release medical information," which served as the cover sheet for the questionnaire itself. As a result, some people, not recognizing it as an official request, threw away the questionnaire, thinking it was "another advertisement or something," or gave it to the children to use as drawing paper.

Next came the problem of completing the twenty-four-page questionnaire, whose format was both complicated and inadequate. The respondents, working without the help of trained interviewers, were to provide demographic data and the names and addresses of their physicians. They were to create a medical history by checking off 148 items and writing in further descriptions of some illnesses on the comment pages. Questions on the comment pages asked whether a diagnosis was established, when, and by whom, including names and addresses of physicians and hospitals. Respondents were to provide a therapeutic history, describing medications and treatments for their entire lifetime; a residential history; an occupational history; a history of exposure to toxic substances; and a social history, with many questions about smoking and drinking. The questionnaire also called for a good deal of history about the adults' families of origin.

Only two pages were devoted to questions about children. These questions, to be answered by females only, were presented on two unnumbered sheets attached at the end of the twenty-two numbered questionnaire pages. The questions about children were birth dates of all children (asked twice); sex and names; dates of all live births, miscarriages, and stillbirths; complications during pregnancies; birth defects; crib deaths; and delivery complications. There were exactly two one-inch-long lines for the replies to "other major illnesses" for all the children in the family. The lines allotted to private physicians' names were twice as long, a fact that did not escape notice. Upon receiving her questionnaire, one resident recalls,

> I threw it on the floor because I was so mad. There was just no place to write about the kids. So they told me that only information about adults

> was important at this time. And that I could write stuff about the kids on the back of the pages. Is that all they cared about sick kids?

She was not alone in her angry reaction to what was seen as indifference to the parents' chief concern. People also noted there was no place to compare patterns of symptoms occurring when people were present at Love Canal or away, or under varying climatic conditions, and no provision for the complexities of emotional problems—simply a place to check for "nervousness." There was good evidence that many people thought the questionnaires confusing and inadequate. Derogatory views were expressed repeatedly—in public and private conversations, at meetings, and in interviews.

The Validity of the Questionnaires

A survey researcher who is faced from the very beginning with a low response rate in a survey can stop to consider the questionnaire instrument itself and the general method of data collection. Assuming that a detailed questionnaire is the way to obtain optimal amounts and quality of data, the researcher may consider other matters. Did the appropriate people receive and complete the questionnaire? Did the subjects understand what the study was about? Were they motivated to respond? Did they have the information requested? Were the questions too technical? Were the questions couched in clear, unambiguous language? Were the respondents able to read and write well enough to complete questionnaires? These are crucial matters, because the answers to major research questions will be no better than the primary data collected. It is well known among survey researchers that the most complete information results when a trained interviewer takes the time to work with each respondent and probe for information about *that* person, not about others (except children). This technique is particularly important in gaining complete information about chronic illnesses, which otherwise tend to be underreported to surveyors (Mechanic and Newton 1965; Elinson 1977). In the Love Canal case, however, this well-known technique was not used. As mentioned earlier, few respondents were interviewed by field workers.

One field worker recalled that they were advised not to worry about answering the people's inquiries about the meanings of questions, but just to be sure the questionnaires were filled in as completely as possible. They were told that DOH people would check up later and phone from Albany. The reason for this advice was that the pressures of the work were overwhelming at that time for the people collecting the data.

If survey questions are not carefully designed to probe accurately into the phenomena of interest and if they are not answered completely, the collected data often will be nothing more than a useless aggregate of numbers and words. In a new situation, the researcher must first explore, sometimes by going into the field for a time and talking with knowledgeable people, often with some of the subjects themselves, in order to gain an understanding of the dimensions of the major problems and to develop hypotheses and general guidelines to aid in the construction of appropriate questions. Frequently, a questionnaire is then refined by trying it out with a small pilot sample before the researcher decides exactly what questions the entire target group will be asked. In the case of physician-researchers, there might even be some exploratory physical examinations or special surveys and discussions with private physicians. These approaches were evidently not followed in the Love Canal case. Rather, the questionnaires were constructed far from the people who were to be studied and were based in part on questionnaires that had been devised for other purposes. We may speculate that one professional rationale for such a course might have been the unthinking assumption that techniques and questions appropriate for other purposes were simply transferable wholesale to this situation. Another might have been the research planners' reluctance to admit that, in this strange situation, questions already on hand and familiar techniques, even if properly used, might not fit the case.

The DOH researchers were also concerned about losing objectivity. The usual definition of research objectivity is that information is derived from the event or phenomenon that is examined, that symptoms are perceptible to others, and that the researcher remains unprejudiced. However, *objective* apparently took a new meaning for the state medical researchers at Love Canal. In the autumn of 1978, for example, one scientist explained objectivity to me: "We deal only with numbers; we're scientists"; that is, scientists must avoid being emotionally swayed in their professional judgments by the sight, smell, and sounds of suffering. This interpretation of objectivity was not confined to the person whose words are quoted here.

Objectivity, however defined, may have been one reason that the DOH made certain choices about the methods it would use to collect data, but there was a territorial reason as well. The DOH wanted to avoid any appearance of intrusion on the existing physican-patient relationships in the area. The DOH was concerned that, if its physicians conducted clinical studies or conducted long, personal talks with individuals, particularly if they gave information to the patients, intraprofessional conflicts would arise. The cooperation of the private physicians was considered so important in the conduct of the health studies that Commissioner Whalen included the matter as one of the recommendations he made in his August 2 order. Even with the order and the DOH precautions, however, relationships between the private and public physicians were uneasy (Powell 3/23/79) and caused problems for many of the residents, increasing their discomfort and sense of mistrust.

Territorial Problems

When the blood samples were analyzed, the DOH sent notes to the residents telling them that the laboratory results would be available from their private physicians. At that time, many local physicians still were not fully aware of the Love Canal sitution or of what the DOH researchers were doing. Many who were aware did not agree that Love Canal presented a particular health problem to their patients. Whereas the DOH probably wanted to avoid sending laboratory reports to lay people that would be meaningless to them without interpretation, the private physicians may have found the sudden appearance of unordered lab results confusing and telephone calls from patients annoying. After the long weeks of waiting for test results for themselves and their children, many residents received cool treatment when they telephoned their doctors' offices. Some were told they would have to pay for office visits to learn about test results or that they would be given test results when the doctor had some time. The private physicians may have had little context themselves for laboratory tests they had not ordered and may have found it difficult to present the findings to their patients in a professionally appropriate way. In any event, some conveyed their annoyance to their patients, who had called at the recommendation of the DOH. The Love Canal people found themselves caught in the middle—hurt, baffled, and angry at both sets of doctors.

Further problems arose when residents asked their doctors to forward medical records to the DOH for the epidemiological study. Some physicians were cooperative about sending the records to the DOH, others agreed but took a long time, and still others insisted on ample payment for copying costs and staff time for the service. The physicians may have resented the requests as impositions on their staff's time. They may have been concerned about the possible use of records in forthcoming litigation. In addition, the confidentiality of a patient's medical records is part of the sacred trust between doctor and patient. Although the New York public health laws protect physicians from action for damages or other relief for furnishing information to the health commissioner on the latter's request (N.Y. Public Health Law, Sect. 206), many physicians may have been concerned about this matter. It was not until a meeting was held between the DOH physicans in charge of the study and the Niagara County Medical Society, and after Dr. Vianna had made personal calls to a number of private physicians, that the impasse was overcome.

Haste and Waste

Although it is true that the DOH had no department specifically prepared to handle exactly the sort of problem Love Canal presented, it is not at all clear why the health studies were treated as a red-hot emergency. The Love Canal situation developed over a thirty-year period. A few weeks' lead time—used, for example, for planning administrative matters, deciding on the best data to collect, designating a comparison group, piloting a questionnaire if one seemed appropriate, becoming familiar with the residents' abilities to provide the information required, training field interviewers thoroughly, assessing laboratory resources, and deciding precisely what information was needed from physician records and how physician records were to be read and coordinated with residents self-reports—might have made a difference, not only in the quality of the data collected but also in the relationship between the residents and the DOH. One can only assume that, under the public pressure to produce immediate answers, the DOH professionals did not demand the time to reflect on what they were going to do, how, why, and what the best use of resources would be before they rushed in and did something.

After the data were collected in the autumn of 1978, the DOH staff spent a good deal of time and effort making repeated telephone calls from Albany to check information with the residents. In all, some 2,700 questionnaires and more than 4,300 blood samples were collected (NYDOH 6/23/80a). It remains to be seen whether the data were collected in a form sufficiently usable and reliable that they can be subject to adequate analysis. The social consequences became apparent very swiftly, however, in the growing mistrust and loss of patience on the residents' part and corresponding defensiveness on the part of the professionals. These sentiments influenced all subsequent interactions between the DOH and the people of Love Canal.

The Swales Hypothesis: The Residents Start Their Own Health Study

A rather remarkable thing happened next, particularly in view of the awe for the professionals' exclusive knowledge that the lay person generally shares with the professional. There was roughly a nine-week period between the decisions of early and mid-August and the beginning of the remedial construction work in October 1978. In that intervening period, the emotional atmosphere for the residents was marked by anxiety, hopes of help, despair that there might be none, and fear of the forthcoming construction's potential for the release of toxic fumes or worse. This period concluded on October 12, when Dr. Vianna's form telegrams came to nineteen families, telling them that their health records did not justify the temporary relocation during the period of remedial construction. Once that occurred, the organized residents mustered their resources and pressed harder for help from the state government. From that day on, rather than depending on the mandates and mercies of the officials and authorities, they took matters into their own hands, essentially challenging the department of health in its special province, although their intent at first had been simply to cooperate with that department.

It all started early in September, when Dr. Vianna asked for the Homeowners Association's help in the DOH survey of families living to the east of the canal. The association workers cooperated vigorously. First, Lois Gibbs met with the organization's newly elected force of one dozen street representatives on September 12. She told them that the decision about "breaking the perimeter"—that is, about further purchases of people's homes and about moving people out temporarily during the construction work—would depend on the health department's researchers learning whether people's health had been affected by chemicals. She told the group: "Dr. Vianna is a good guy. If you talk to him when he is up there on stage he'll tell you what sounds right in front of the state people, but you get him aside—he's a good guy—he's for the people!" Thus, she enlisted the help of the representatives in urging people on their streets to complete the health questionnaires, to sign release forms, and to ask physicians to forward records to the DOH in Albany.

The street representatives notified their people, but, when that effort produced inadequate results, the handful of most dedicated association volunteers decided to telephone every family to urge cooperation with the DOH study. They believed the work was important, and they cheered each other on through hours of tedious telephone calls. It was then, as mentioned earlier, that they learned that many people were confused. Some people did not understand the importance of signing the releases for doctors' records and of filling out the questionnaires completely; some thought they had already signed the necessary forms when they signed an authorization on the questionnaire itself, releasing the information in that instrument for scientific purposes. Many had not yet signed the releases because it required a trip to the association office to pick up the release and then another trip to the physician's office to leave it with the secretary or nurse.

In the course of the rather lengthy telephone conversations, the residents not only became fully aware of the association, but many also told their callers about their troubles, including their medical problems. At first, the women making the calls simply kept track of who had been called, but then, as they talked, they began to jot down a rough record, with brief notes about the medical information volunteered during the conversations.

One evening, in the third week in September, Lois Gibbs was looking over her materials to prepare a report for Dr. Vianna, who was to be in town the following day. By then she was becoming concerned that it might take years to establish whether there were health problems at Love Canal. She had also heard disquieting rumors that the governor intended to purchase no more homes. A task force news bulletin of September 15 clearly stated that the construction project to begin on October 10 was to be performed with such safety precautions that no plan was *necessary* for evacuating residents but that, as a concession to the residents' concerns, "A great deal of time and effort has been expended on the development of the Evacuation Plan by experts" (NYDOT 9/15/78). If no temporary evacuation plan were even deemed necessary, then Gibbs's hopes for any further long-term relocations were beginning to fade.

Feeling rather desparate, because she wanted out for herself, for her family, and for all the residents who felt as she did, she worked throughout the night at her kitchen table, poring over the telephone-call records spread out before her. She began to place pins on a street map, to mark the illnesses reported to the association's telephone callers. It seemed to her that patterns of illnesses emerged, marked by a few rough clusters of pins here and there and some wavering lines of pins as well. As she examined these patterns, she recalled that, when she had trudged from door to door in the summer, she had noted several times that people who were neighbors told her about similar health problems.

Because Gibbs was familiar with the area and with the people's reported problems, she could use additional information known by others but not yet considered related to health. At a public meeting in August, Dr. Kim had mentioned that the DOH might use aerial photographs dating back to the 1930s to locate old drainage areas, and that they might take soil and air samples from homes in those locations. On September 15, an information bulletin from the task force mentioned that there might be underground water courses and provided a speculative map of their location (NYDOT 9/15/78; Adams 12/20/78). Most important to Gibbs, some of the older residents had come to the association headquarters and had talked about huge swales cutting through Love Canal and running for long distances through the area. They had reported the presence of these now-covered-over ditches to the task force as well. Swales are natural drainageways that can provide preferential routes for the movement of liquids underground. Gibbs became very excited when she suddenly realized that chemical leachates might be traveling through the underground swales and that illnesses related to the chemicals might show up in clusters and lines cutting across streets, representing the swale paths and possibly underground ponds as well.

The next day, Gibbs met with Vianna and showed him a street map with colored circles, triangles, and squares pasted on it to indicate reported ill-

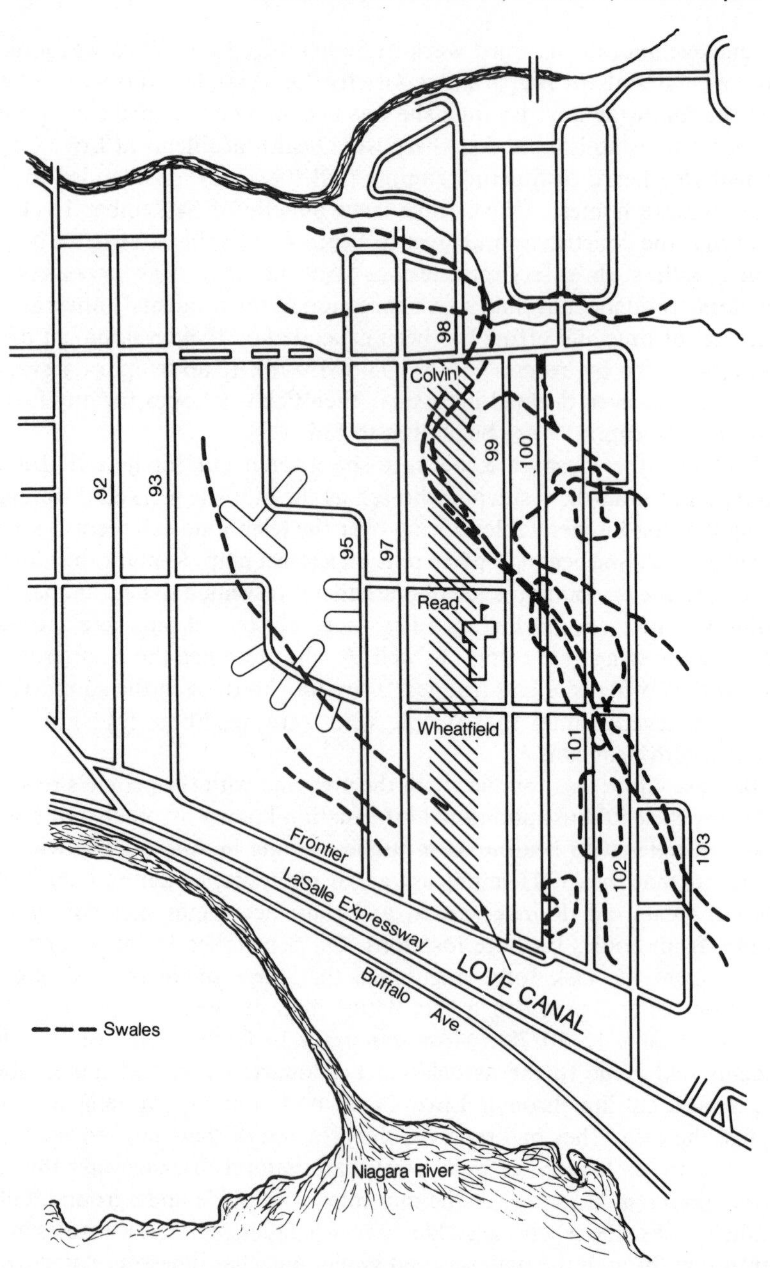

Note: The 99th Street School building is located between Read and Wheatfield. The LaSalle Housing Project (Griffon Manor) is between 93rd and 95th Streets.

Source: Map was drawn by CWP Graphics, Washington, D.C..

Figure 5-1. Schematic Map of Love Canal Neighborhood

nesses. She told him eagerly that there might be a relationship between certain health problems and the location of the old streams. He was calm and told her that, if she could specify the location of the swales, he would "plug it into his computer."

Somewhat encouraged by his remark, Gibbs called Dr. Beverly Paigen and asked her for help. Paigen's two major research interests were detecting the presence of mutagenic chemicals in aquatic sediment and genetic susceptibility to environmental toxins. Her interests had attracted her to the Love Canal area as a possible research site for studying family patterns of genetic reactions to chemicals. An environmental activist, she was also interested in developing efficient, inexpensive, easy, painless ways to assess health hazards from chemical wastes. When Wayne Hadley had alerted her to the studies undertaken by the DOH in the late spring and early summer of 1978, she immediately wanted to get a clear idea of what the DOH people were doing. Her goal was to conduct research at Love Canal that would be useful in her research program and would complement the work being done by the DOH. She had become aware of the residents' needs for information about matters of crucial import to them. For all these reasons, Paigen was very willing to help when Gibbs turned to her.

At this time in September, Gibbs and other residents were recognizing their need for outside experts as they began to doubt the wisdom, competence, and motives of the state agencies. They were also learning that, without money, scientific experts were hard to find, notwithstanding the professional ideology of doing work for its own sake or for service only. By now, the residents had been promised that the state would hire a toxicologist expert in environmental studies, who would report to the residents as their consultant, but he had not yet arrived. The residents had no health consultant other than their own physicians and the DOH. Thus, they were overwhelmingly grateful that someone of Paigen's competence, scholarly credentials, and interpersonal skills would take a professional interest in the situation.

Paigen was intrigued by the information Gibbs gave her and excited by the implications. She realized that, if there were some relationship between the paths of underground drainageways and illnesses, then examining samples from ever-larger concentric rectangles leading outward from the canal—in the pattern that had been part of the DOH researchers' original plans—would not detect such relationships. Paigen decided to conduct a study that might be helpful to the DOH researchers. As most survey interviewers are lay people trained for the job they are to do, Paigen believed that the eager volunteers in the Homeowners Association were fully capable of the task. She told Gibbs and her coworkers to start over, and she helped the women devise a set of questions and a procedure that would be scientifically acceptable.

> They put aside all the information they had gathered and started making a systematic phone survey to each home, collecting information about the number of persons in each family, the length of time they had lived in the Love Canal area and the health problems experienced by the family. More than 75 percent of the homes cooperated in the survey. (Paigen 3/21/79, p. 12).

Each respondent was asked whether the illnesses described had been treated by a physician or in a hospital. Those illnesses not so treated were listed separately but not included in the compilation of the data that formed the basis for reports by Paigen to the DOH, to the residents, and, later, in her testimony before a congressional committee (USHR 3/21/79).

The callers did not ask for medical records to substantiate the verbal reports. As lay people, they could not very well request such records; furthermore, because the DOH was collecting the records, Paigen assumed that, if the findings indicated any relationships, the DOH researchers would simply check their far more complete data against her findings. This data collection began in late September and continued for about three and a half weeks.

In early November, at her insistence, Dr. Paigen shared the results of the study with the DOH researchers. Accompanied by Stephen Lester (the newly hired toxicology consultant to the Homeowners Association), Paigen went to Albany—numbers, tables, and maps in hand—to describe the partially analyzed data.

The Homeowners Association workers had telephoned all residents on the three streets east of the inner ring. Paigen had information on 594 children and adults living south of Colvin Boulevard (that is, within the defined Love Canal neighborhood) and on 546 people from an area north of but contiguous with the defined Love Canal area. The northern area could serve as a control group, but only a rough control, because those in the northern areas could only be considered relatively less affected but not unexposed to the hazards of chemical wastes, given the uncertain boundaries of the canal and the drainage maze extending through the area. There were wet-area homes (built on swales or former swamps) and dry-area homes in both sections. Of the people included in this survey, 236 of the 546 from the northern group and 78 of the 594 from the southern group were from homes in wet areas. The data may have underestimated the true incidence of illnessess, because they did not include the 239 presumably most exposed families who had lived in inner-ring homes. Furthermore, the people who supplied information might have been reluctant to give complete information to lay people from their own community of matters such as nervous breakdowns, siezures, and some other disorders.

Although her formal research training was not in epidemiology, Paigen was an experienced scientist, and she presented what she regarded as an imperfect set of data, with the limitations clearly spelled out. Her purpose was to alert the DOH researchers to suggestive information on the distribution and frequency of reported miscarriages, kidney and bladder disorders, and central nervous system problems. She had concluded that there might be relationships between wet versus dry geographic locations and reports of disease and miscarriages, and that more focused studies should be done by trained DOH professionals (Paigen 10/31/78; Shribman and MacClennan 11/1/78).

Paigen was pleased with the reception she thought her suggestions received at her November 1 meeting in Albany with the DOH scientists. She was not surprised at the reception, for she assumed she had a comfortable, mutually respectful working relationship with the DOH scientists working at Love Canal. At the meeting, they seemed interested in the scientific implications of her careful presentation. However, the mutual goodwill of the scientists, and their potential for working together, was marred by the political and economic context in which their discussion took place. Paigen had already experienced Commissioner Whalen's displeasure at her interest in Love Canal (to be described in the next chapter), but she had dismissed that as a top-level administrator's stuffy fear that a younger colleague might somehow rock the boat a bit too publicly by showing that scientists disagree with each other. She believed that controversy is the very stuff of science, and that the methods scientists had developed over the years would be sufficient to handle scientific disagreements.

The study had already been publicly attacked, however, before Paigen made her presentation at the Albany meeting. In the wake of the angry disappointment of the residents who had received, or had heard about, the form telegrams sent by the DOH in the second week of October (with the curt notice that health records indicated no reason to move residents from the Love Canal area during the period of construction), Gibbs, without consulting Paigen, had released some of the early results of the Paigen-Homeowners Association study to journalists (Brown 10/4/78). This was certainly not the usual procedure for release of a scientific study, prior to peer review or other scientific safeguards. Dr. Vianna's response to inquisitive reporters was that the data in the association study were "totally, absolutely and emphatically incorrect" (Shribman and MacClennan 10/18/78)—certainly not the usual procedure for describing data as yet unseen and unevaluated. (It was later reported that DOH officials characterized the evidence as "information collected by housewives that is useless" (MacClennan 1/5/79)).

On the weekend prior to the scientists' November 1 meeting in Albany, a local newspaper featured a front-page story quoting Lois Gibbs. She charged that the health department was minimizing the severity of the residents' health problems, and she threatened that a citizen demonstration would be held on that weekend when President Carter arrived at the Buffalo airport to campaign for Governor Carey's reelection. President Carter had not answered her repeated telegrams, the governor had reneged on his earlier promises and was not about to consider futher temporary or permanent moves, and the health department seemed unresponsive to residents. In general, the residents felt they had good reasons to vent their feelings of frustration (Shribman and MacClennan 10/28/78). On November 1, while the Albany meeting between the association's consultants and the DOH scientists was taking place, another news article, illustrated with a simplified map, proclaimed that there might be more confirmable illnesses at Love Canal than the state had yet studied—the very point Paigen was making to her peers (Shribman and MacClennan 11/1/78).

The Albany meeting took place three days after the airport demonstration. Later Gibbs was told privately that Governor Carey had been made very uncomfortable by the sight of the demonstrators at the airport, who appeared ungrateful for the tremendous efforts already made on behalf of the Love Canal neighborhood. President Carter's first sight of the Governor in his own bailiwick was flawed by the presence of disgruntled people holding huge signs pleading for even more help than the millions already committed from the state's coffers. When President Carter told Debbie Cerrillo, the association vice-president, that he would pray for the residents, the governor was evidently no more mollified by this promise than was Cerrillo.

The word in Albany soon was that "those people in Western New York aren't going to get another cent." It is not known whether this statement was completely accurate, whether the governor's anger was shared by the DOH scientists, whether they thought that Paigen had colluded with Gibbs to release the results of the study in the newspapers in the attempt to put pressure on them to recommend further moves, whether they were insulted by what they regarded as an attack or felt betrayed in some way by Paigen. However, Paigen's suggestions and the Homeowners Association findings were rejected firmly at that time, although that was not to be the end of the little study, and we will return to it more than once.

The next matter of importance at Love Canal in the autumn of 1978, as far as the residents' health and futures were concerned, was the beginning of the analysis of data that were still being collected by the department of health.

Mistrust and Miscommunication: Data Analysis Begins

As it became clear to them that all the remedial, helping actions that would be undertaken were already firmly decided as far as the state authorities were concerned, the Love Canal residents listened suspiciously to the DOH officials' cautious statements regarding the analysis of both environmental and health data (see Brown 11/14/78). For example, reports had appeared in the news that dioxin had been located in the area. When DOH officials, in their efforts to stick to fully confirmed facts, asserted that they had no firm evidence of the presence of dioxin in the Love Canal leachates and that, if present, the dioxin might not migrate because it is not soluble in water (Brown 11/8/78; 11/9/78; 11/10/78; *Buffalo Courier Express* 11/10/78; Spencer 11/11/78; Shribman 11/12/78), the Love Canal residents suspected that the pronouncements emanating from Albany were part of an organized attempt to minimize the danger that the residents might be in and thus mask the need for further expenditures on their behalf.

Dioxin is the common name for 2, 3, 7, 8 Tetrachlorodibenzo (p) dioxin (TCDD), a particularly toxic contaminant produced in the synthesis of the herbicide trichlorophenol (see World Health Organization 6/78). It is the substance responsible for the highly publicized disaster at Seveso, Italy, in 1976 (Whiteside 9/4/77, 1979), in which hundreds of people were forced out of their homes for months. The biological effects of the substance are said to be apparent in concentrations of parts per trillion. The very word *dioxin* frightened people, and some residents found a popular version of the Seveso story and passed it on to their neighbors (Fuller 1977).

At this time, while the DOH was cautiously analyzing data, Lois Gibbs was under pressure from some of the more impatient members of her organization, who were threatening "to take some *real* action." She repeatedly attempted to persuade both the head of the task force and various DOH officials to station knowledgeable health authorities at the Niagara Falls site to meet with people regularly and, preferably on an *ad hoc,* informal basis, to respond to their many questions; if not that, to hold regular formal meetings; and, at the very least, to provide her with correct information with which she could answer the inquiries she received day and night, from anxious residents. The questions concerned dioxin, chemical air readings, soil samples, construction work, blood-test results, and many other technical matters. Gibbs knew how stoically the people reacted to solid information—making realistic plans for regular examinations, special diets, moves, and changes of all sorts. She also knew that the community members became angry and frustrated when their need for information was ignored or evaded by scientists and officials, who seemed to imply that they knew the answers but just were not telling, or that they did not know the answers and were not eager to find them.

In the increasingly tense atmosphere created by the sheer lack of contact between the DOH and the residents, the occasional official reiteration of the scientifically conservative position of waiting for absolute confirmation of cause-and-effect links before releasing any results, the public criticism of the DOH by association leaders and other residents, the release of the association's health study findings to the newspapers, and the reports and denials of the presence and effects of dioxin, a meeting was finally called by the department of health to present a progress report on the results of the health and environmental studies. On November 21, 1978, Dr. Glenn Haughie and Dr. Robert Huffaker (the state's on-site safety officer for the remedial construction) conducted a very heated meeting at the 99th Street School auditorium, which was overflowing with residents. Several important matters were revealed at this meeting. First, Dr. Haughie conceded that there was evidence of chemical contamination of soil along the former canal's drainageways but said there was no evidence that people living along the former swales had unusually high rates of illnesses. He did not say whether the proposition had been tested and found wanting, or whether "no evidence" simply meant "untested as yet."

The second issue had to do with identification of the chemical compounds found in the leachates from the canal. Dr. Haughie announced that more than two hundred compounds had been identified, and he started to illustrate his words with slides listing the compounds. The local newspapers' accounts of the possible presence of dioxin at the canal site were well known to residents, who were grasping for every bit of information available from any source. Now their fearful questions about the presence of dioxin were turned away with the response, "no evidence." The audience became so angry and noisy that the slide show ended abruptly, and the meeting was taken over by residents' questions. Whereas the health department officials might have meant that it is extremely difficult to test for the minute quantities of dioxin that can be toxic to human beings, and that such tests might *not* have been performed or analyzed at the time, the citizens treated their answers as a bald attempt to deny that dioxin was present at all. Often, the phrase "no evidence" has been interpreted in the public press, and even by the educated lay and scientific public interested in Love Canal, as meaning "All the tests *have* been performed and there is no evidence," a very different meaning indeed from "No evidence because no tests have been performed," or "The tests have been performed and no *statistically significant* differences have been found." The DOH frequently failed to distinguish these meanings or clarify exactly what they had done and had found.

A third matter was the news that general liver function tests performed on the blood samples indicated liver abnormalities among a small number of residents of the first ring. Dr. Haughie did not discuss in this heated public meeting that, in their first analysis, the DOH researchers thought that there were abnormal findings on the liver-function tests for about 13 percent of the first-ring residents. However, after they made corrections for age and sex, their sample showed no apparent excess of abnormal findings when compared with a report based on a study of hospitalized patients. The liver-function analyses and adjustments had been explained to Paigen and

Lester in a confidential meeting in Albany on November 16, but the data on which these conclusions were based were not supplied to Dr. Paigen, despite her repeated requests (Paigen 3/29/79).

The information about the corrections to the liver-data analysis had been conveyed to Lois Gibbs by the two consultants. She then faced the dilemma of whether to keep quiet about information that was relevant to the health of the residents. She remained discreet about the analysis for some time—until she began to war openly with the DOH.

At the November 21 meeting, Dr. Haughie stated that some young boys who lived in wet areas showed signs of abnormal liver functioning. On the evening prior to the meeting with the residents, the local physicians had been informed about the problem and about a planned follow-up examination of some fifteen boys under age nine. The DOH had not spoken directly with the parents of the children, however, to explain what they planned. As a consequence, once again, needless fears and mistrust were aroused. As for the dioxin rumors, the substance was finally confirmed to be present at Love Canal in huge quantities, far greater than had been reported at Seveso, Italy (Brown 12/9/78; MacClennan 12/20/78). The confirmation itself, and the way it was done, once again increased the residents' mistrust of the health department officials' good intentions and capabilities.

Answering Questions

For a number of reasons—including the cryptic health study reports, the lack of DOH communication, the poor adherence to safety standards by the work crews at the construction site (Clement Associates 8/20/80), and the general feelings of frustration among some residents—the organized residents decided to mount a picketing action to slow or halt the remedial construction project. On December 8, 1978, at the pre-dawn hour when the construction workers arrived, residents started to picket. They marched from 5:30 to 9:00 a.m. every day, back and forth in front of the tall gates in the fence surrounding the construction work. During the second week of picketing, on a freezing morning when six residents were arrested, a sudden meeting with the health department was announced. The residents' excitement quickly turned to disappointment when the health official simply read a news release from the health commissioner in a slow, mechanical drone. The news release stated that dioxin was present in the canal area and that, although the health department had suspected its presence all along, there was no evidence that the amounts found posed an immediate health hazard to the residents (NYDOH 12/11/78). Once again, the residents viewed the health department's conservative position as an evasion of a clear and present danger. To questions about the blood tests and soil samples, the official repeatedly said, "I don't know," "I can't answer that," or "I am not a toxicologist." The exhausted picketers were smoldering with resentment that the confirmed presence of dioxin was described to them by an uninformed individual and by a release prepared for the newspapers rather than by direct communication from knowledgeable DOH officials to each of them or to their organization's leaders. Months later, they learned from reporter Michael Brown's detailed description that Dr. David Axelrod had telephoned him with the news of the dioxin findings on Saturday evening, December 9, a few days before the residents were notified (Brown 1979).

The person who conducted the meeting on that cold, gloomy morning in December mentioned later that he felt proud he had been able to remain cool, "to talk like a machine," despite the anger the Love Canal residents displayed when he read the DOH announcement. Task force members expressed similar feelings more than once when they conducted or even were present at meetings with residents who seemed so unreasonable. Privately, the officials congratulated each other on not giving anything away, on not conceding anything to the residents. In public, they said they saw "no evidence," or said "I don't know," and argued that the people were given answers, but, "like spoiled children," just did not like the answers they had received.

There were two issues here. Even if sharing information had been seen as truly important, it was difficult to communicate in this emotionally charged situation. The very act of talking with the people had become an ordeal, especially in sporadic, large public meetings. Marathon confontations became grueling events to be endured as part of an already tough job, as accomplishments to share with colleagues who had also earned their "combat scars." Simply living through a heated meeting, getting it over and done, without "losing cool" and without departing from an official position became one more mark of the professional.

The second issue concerned exactly why the questions were unanswerable. First, the answers to many questions are even now not available, because the scientific knowledge does not exist. In fact, it is from proper studies of such tragedies as Love Canal that some answers could possibly emerge—a fact of but small comfort to current sufferers. There were no answers to other questions, however, because of apparent deficiencies in the very areas in which the DOH personnel were supposedly proficient: collecting and analyzing health data in scientifically acceptable fashion.

To the implications and outright accusations of bungling, the health department personnel responded evasively and defensively. The attitude that the professional knows best can be sustained with dignity, however, only when the professional exhibits skills above and beyond those of the lay person. Otherwise, the claim that there are realms of esoteric knowledge open only to the professional appears ludicrous. "I don't know," which may, in fact, be a perfectly correct response even now, given the state of knowledge, thus appeared to indicate the lack of worth of the very people in positions to make crucial decisions about the Love Canal residents' lives.

The next important events that took place eroded the Love Canal people's trust even further. This time, rather tragically, the source was the DOH person the residents respected and trusted above anyone else, David Axelrod.

The New Commissioner Learns His Role

In the middle of December 1978, Governor Carey announced the appointment of Dr. David Axelrod to replace Dr. Robert Whalen as commissioner of health. Axelrod, director of the health department's division of laboratories, was widely known for his interest in environmental toxicology. In January, soon after taking office, the new commissioner announced several steps that were viewed positively by the residents who were still striving to have their cases considered for relocation. He confirmed that the health investigations would be extended in order to determine whether chemical migration had proceeded further than had been envisioned in early August. Furthermore, he said that data from the health department studies were to be ready for use by January 15, 1979, when the materials would be in the computer (MacClennan 1/5/79).

A group of worried mothers surrounded Dr. Axelrod when he came to talk with other state health department personnel in Buffalo in mid-January. He assured them that he felt confident that the ongoing construction would take care of the long-range problems in Love Canal. He was gracious to Patti Grenzy, a young woman who was two months along in her pregnancy the previous August and had been living in fear ever since. Her front windows faced the high fence some twenty feet from her door, separating her home from those purchased by the State. Dr. Axelrod told her kindly that he was sure that her unborn child was in no more danger than any other unborn baby anywhere. (see Fergus, 1/12/79) He answered and returned phone calls. Dr. Axelrod's behavior, attitudes, and demeanor seemed warm, kindly, and attuned to the residents' needs.

In mid-January, Dr. Axelrod invited the residents' consultants, Stephen Lester and Beverly Paigen, to attend the first of two meetings he conducted with a "blue-ribbon panel," whose names, the commissioner cautioned, were not to be made public. The panel's purpose was to review the department of health's findings. A public meeting was called for February 8, and the blue-ribbon panel met again, without Lester and Paigen, on February 7 (Brown 1/23/79; MacClennan 1/29/79). Hope was rekindled among the residents, who had been feeling abandoned and in despair as the last of their inner-ring neighbors left and as the Task Force staff was systematically reduced.

A few days before the public meeting, the Homeowners Association conducted a brief survey by telephoning 101 inner-ring families who had been relocated for several weeks to a few months (the only follow-up health study performed by anyone as of Autumn, 1981). Sixty-seven of the 101 respondents said that the health of their family members had improved dramatically since they had left the canal area. The relocated residents did not simply report improved health in general terms. Rather, to the question "How do you all feel since you have moved?" they specified a decrease in or disappearance of specific symptoms since moving. Thirty-two replied there had been no change, and two felt worse. The association duly reported their additional findings to the health department during the first week of February in the hope of showing that there were, indeed, illnesses related to residing at Love Canal.

Tension and excitement mounted among the Love Canal neighbors as the meeting day approached. On February 8, the day after the DOH-blue-ribbon panel meeting, there were endless speculative discussions among the residents, focusing on possible decisions to be related to them that evening.

As had become the custom, however, the decisions had already been reported to the world well before Dr. Axelrod read his supplemental order to the residents. In an interview on the CBS morning news on February 8, Dr. Vianna stated that the sole illnesses found in the Love Canal residents were higher than normal rates of "fetal wastage" (miscarriages) low birth weights and birth defects. He did not mention that only these reproductive factors and the signs of liver damage in a group of young boys and men had been studied at all (so far as it is possible to discern). He did not mention that there were no comparison groups for the massive study. He also stated that the inner-ring homes had been bought to facilitate the remedial construction, not because of health problems.

This was the first time residents heard the term "fetal wastage." The parents of these "products" had been accustomed to thinking of them as "the baby we lost". "Adverse pregnancy outcomes," they were to learn, included low birth weights, ("tiny Julie"), birth defects, ("our baby with the clubfoot") and stillbirths ("the other baby we lost"). Although the terms are technically correct, their use of these terms, helped to increase the sense of distance between the DOH and the residents.[2]

Those who missed the morning news could read the full set of recommendations in the newspapers before the scheduled meeting with Dr. Axelrod.

The Relocation of Additional Families

The February 8 meeting was packed with people, blinking under the glaring lights of television cameras ready to record the interchanges between residents and an array of state and local officials. The speaker everyone was eager to hear, however, was Dr. Axelrod.

Axelrod graciously acknowledged that the Paigen-Homeowners Association study had alerted the DOH to the significance of the "historically wet" area. Then, in general terms, he described his meeting with the nameless, expert blue-ribbon panel members. He said that he and other DOH scientists had presented three sets of evidence to the blue-ribbon panel on the previous day. The information was about "fetal wastage," fetal malformation, and babies with low birth weights. Fetuses were the most important indicators, he explained, because they are the human organisms least resistant to environmental stresses.

Dr. Axelrod said there were twofold increases in the three indices among the residents of wet areas. He called these increases a "small but significant risk for fetuses." Finally, he said that he had recommended to the governor that pregnant women be moved from the area at the state's expense, if they wished to move. Although he said there were no demonstrated clinical effects on young children, he recommended that children under age two should also be moved as an extra health precaution.

On hearing this, people began to shout, many with tears streaming down their faces. One resident cried, "My child was under two last August. It's your fault we didn't leave then!" A man whose wife had suffered several miscarriages shouted "You aren't even human! Humans couldn't do this to each other! You're just trying to pacify us." And the familiar cry went up, "We want *OUT*!"

Some comic relief was provided by a man who arose, hand in hand with his wife, and shouted as they walked up the aisle, "I'm going home to get my wife pregnant so we can get out." The audience laughed but it was not a joyful sound.

The Tools of Science and Their Uses

This meeting had an agitated atmosphere, with a sense of deep, barely checked antagonism toward the state officials. Later, there were a few moments of calm, when Dr. Beverly Paigen made some spontaneous comments (to be discussed in the next chapter). A few older members of the audience left during the most heated exchanges—some fearful, some put off by the disrespectful remarks and questions flung at the commissioner. As the meeting went on, Commissioner Axelrod walked away from the podium a few times when the comments became too insulting, but in general he tried to speak reasonably and fully to everyone. He asserted that he had not made any *decision*, only the *recommendation*, and that the governor made the decision. The nonscientists living at Love Canal did not grasp that fine division of labor between the commissioner of health and the governor.

The outer-ring residents had long thought the evidence showed that contamination and health hazards had spread well beyond the arbitrarily determined perimeter. The health commissioner, however, asserted the DOH's continuing goal of trying to confirm *absolutely* that health effects were related to chemicals and not to any other factors. Health department scientists, for example, pointed out the necessity of eliminating smoking, drinking, and other occupational hazards as possible causes for health disorders; residents countered with examples of illnesses in young children.

His recommendations, Commissioner Axelrod asserted to the angry audience, were in the "conservative," health-preserving direction, given the findings so far. He stressed the importance of validating data. When asked by an observer, he stated that the DOH was using a confidence level of .05.[3] The audience was not knowledgeable enough about statistical jargon to know that the "confidence level of point oh five" is simply a convention adopted by scientists (see Blalock 1960), not at all immutable, particularly when decisions must be made concerning people's lives and well-being. They did know, however, that there was no precise information about the location of the chemical leachates, nor about the potential health effects of every chemical compound and mixture of compounds. They also were keenly aware that they were living in or very near an area already declared hazardous to health. They thought the decision problems for the authorities were rather simple and clear: should a great deal of money be spent to move everyone out, if even slight evidence of serious harm existed, to make certain that no further chemical-related damage could occur to their health; or should there be absolute certainty that serious damage had occurred, with the nature and source of damage known, before the public monies were committed. "What does it take," cried one woman, "asthma, three miscarriages, a birth defect, a man with a damaged liver! You're a doctor! You should care about us!"

Sharply attuned to the problems of costs, as a result of their own daily skirmishes with the realities of stretching low incomes to cover normal living expenses, the residents reasoned that the high-level health decisions depended on which way it might cost more to guess wrong and on who would pay the price for any decisions, right or wrong. If the choice were made to move people out before incontrovertible evidence of health damage was available, then there was a good chance of spending a great deal of money needlessly. With the choice of "wait and see" or "the chemicals are innocent until proven guilty beyond the shadow of a doubt," money would be saved in the short run; but later, the residents' health might turn out to have been damaged. Since *their* health and *their* financial futures were at stake, and since they were taxpayers and citizens, the residents thought that not only the decision problem but the solution was absolutely obvious: take the chance on spending the money.[4]

This type of decision is apparently not clear and simple for scientists working in real-world situations. Scientific truth is supposed to be sought by people who are unswayed by emotional, monetary, or other considerations. Scientists who are called to participate in policy decisions have developed the rationale that, as scientists, they will confine themselves to assessing the probability of events happening and let *other* people decide what to do about it. For example, a May 12, 1978, DOH memorandum discussing "acceptable involuntary risk" states that "we look to scientists, technologists, epidemiologists and others for an objective evaluation of hazard," while "safety . . . is a political judgment" (Axelrod 5/12/78).

Judge David L. Bazelon has commented:

> In reaction to the public's often emotional response to risk, scientists are tempted to disguise controversial value decisions in the cloak of scientific objectivity, obscuring those decisions from political accountability. (Bazelon 1979)

The Love Canal residents put it even more simply. They pierced right through the cloak of scientific objectivity. On that night in February, and at other times as well, they repeatedly asked officials: "Would you live here?" "Would you bring your family here?" "Would you trade houses with me?" In short, they went to the heart of the matter: acceptable levels of risk?—acceptable to whom? And who, they wanted to know, was to bear the heavy burden of the risk?

The meaning and policy implications of using the fetus as an indicator of chemicals effects were viewed very differently by the Love Canal families and by the health officials. Commissioner Axelrod insisted that, when he recommended that both children under age two *and* pregnant women move, he was offering protection well beyond the essential protection of fetuses. A different interpretation, favored by the residents and their scientific advisors, was that the fetus, the most sensitive human organism, acts much as a thermometer or other measuring device—showing that a danger is present, affecting all but showing up only in the indicator at first.

There was little or no argument about the time at which the fetus was in greatest danger from the chemicals in the mother's body—during the first trimester. With the bureaucratic apparatus constructed by the task force to implement the commissioner's recommendation, however, by the time a woman suspected her pregnancy, had it confirmed by a physician, completed applications, received approval for a proposed move, and located a temporary residence, the families feared that the period of maximum danger to the fetus had already occurred.

Although the people wanted information about their health and wanted that information to be as correct as possible, they were also fully aware that the insistence on unassailable standards of scientific proof fitted smoothly with the political decisions not to move people, even temporarily, while the massive construction project was going on. On the day after the tumultous meeting of February 8, any shreds of belief people might have maintained that the decisions were based exclusively on scientific data were severely undercut. They read in the newspapers that the relocation recommendation had been changed to include people in the area to the *west* of the inner rings, the area including the tenants in the public housing project, for whom no data had been analyzed (Shribman and Johnston 2/9/79). The inclusion of this group in the relocation recommendation was interpreted by the residents as one more confirmation that political pressures were ruling the decisions about their lives.

Within a week, Gibbs and several other Homeowners Association leaders made a well-publicized trip to Albany, carrying mock but realistic-looking child-sized coffins festooned with blue ribbons and slogans—to present to the Governor. Gibbs charged that the newest decisions were political, not scientific at all (Shribman 2/15/79). When the governor's aide, Jeffrey Sachs, tried to argue that the health statistics in the South Bronx were worse than those at Love Canal, and that the organized citizens should be pressuring the federal government for more funding to help the state, (Ackerman 2/16/79), such statements only increased the people's distrust of the scientific rationales for the decision to leave most of them in their homes.

During the months following the February 8 order, more than forty families moved to temporary quarters under the program whereby the state provided funds so that the families would pay only their normal monthly housing expenses. However, Dr. Axelrod received frequent telephone calls from young women who were not yet pregnant but who wanted to have more children and were concerned about the risks of miscarriages, birth defects, and low birth weights. Although he had no specific answers, Dr. Axelrod tried to be reassuring. On one occasion, he even joked with a caller, telling her that, at thirty-one, she was still a young woman and could still have many more children. She responded, furiously, that she hoped for his sake that he and his wife never suffered the torment of doubts and fears she and her husband were undergoing.

As before, these responses were widely discussed in the neighborhood and frequently publicized. Some of the women tried to reach Dr. Axelrod in a personal way as well. One woman sent him a Father's Day card bearing the names of women who suffered miscarriages. Letters were sent to him listing the deaths in the area. In these and other ways, the residents tried to make the commissioner truly feel their plight, feel the hurt as a fellow human being.

If we believe in the strong influence of the situation on the individual, then we must acknowledge the great pressures and conflicts inherent in the commissioner's position as a political appointee, occupying the "hottest seat" in the first place in the country where a public health problem of enormous dimensions had surfaced. He was saddled with the decisions, procedures, and behavior of predecessors and bureaucratic colleagues. One can see the difficulties inherent in the role when, for example, his "extra precautious" recommendations for moving more people out were received ungratefully by people who publicly reviled him. Within a few weeks of assuming office, the commissioner underwent a sort of baptism of fire. In the weeks and months to come, the very man who had played such an instrumental role in discerning the public health problem became one of the principal villains in the minds of Love Canal residents.

Summer 1979: The August 21 Meeting

One final public meeting took place between the residents and the commissioner of health. At that meeting, on August 21, 1979, the chasm between the two sides—the residents and the DOH—widened beyond repair. During the long hot days of June, July, and early August, when the huge remedial construction project was going full tilt for long hours, the chemical stench hung heavy in the humid air. It was months after the DOH had completed their collection of health data, well after a series of statements by the governor that no more people would be moved from the area (Allan 2/18/79; Shribman 2/22/79) and that people could not be protected from all risks (Brydges 4/27/79). The latter statement had also been reiterated by the health commissioner.

Because their requests for the results of the health and environmental studies went unanswered, residents were becoming convinced that there was a cover-up of the true conditions of their health (Gibbs 8/9/79). In the months after the February 8 order, the cold anger between Commissioner Axelrod and Lois Gibbs increased. Gibbs felt she represented residents whose fear and anxiety increased every day, while they were literally trapped in a place whose health hazards had been declared but not specified to them. The actions Gibbs took must have appeared to the commissioner as badgering, public embarrassment, and the expression of lay opinions on matters that could only be understood by scientific professionals.

Shortly after the February 8 report was issued, Beverly Paigen followed one of Gibbs's longstanding suggestions that the DOH information on miscarriages, low birth weights, and birth defects might be better understood if it were examined in five-year intervals, to see whether changes had taken place over time. Paigen wrote to Dr. Vianna requesting this sort of analysis (Paigen 2/16/79). One month later, Dr. Haughie replied:

> Concerning your notion about examining pregnancy data by five year time intervals, it is important to note that during the 35-40 year period during which pregnancies were counted, there were only 80 live-births and 25 miscarriages, 11 birth defects and 13 infants with low birth weights. [Among residents of wet-area homes, east of the canal, south of Colvin Boulevard]. As you can appreciate, distributing these rather small numbers over seven or eight time intervals makes difficult interpretation. (Haughie 3/15/79).

Even the "uneducated housewives" of Love Canal were well able to come up with the average of a 24 percent chance of a miscarriage (25/105) and a 36 percent to 46 percent chance of miscarriage or birth defect or low birth weight (38/105 to 49/105). Over the spring and summer, Gibbs continued to call the commissioner regularly to tell him the troubles the residents were having as a consequence of the construction work. She voiced complaints publicly, spoke at public gatherings of people interested in environmental problems, attracted mass-media attention as frequently as she possibly could, and wrote a letter to the editor of *The New York Times* (Gibbs 8/9/79). She also testified at the invitation of a congressional subcommittee in March 1979 outlining in detail what she and other residents viewed as deficiencies in the DOH handling of the Love Canal case (USHR 3/21/79, pp. 87-92). The association's scientific adviser, Beverly Paigen also complied with the congressional subcommittee's request to testify. Under oath, she presented the results of the study done with the data collected by the Homeowners Association, complete with tables, maps, and written analysis (USHR 3/21/79, pp. 60-69).

On July 30, 1979, Commissioner Axelrod sent an interim report on the DOH environmental and epidemiological studies at Love Canal to Commissioner Hennessy (Axelrod 7/30/79). Hennessy then sent it to members of the task force and to the citizens organization leaders (Hennessy 8/6/79). Lois Gibbs countered within a few days with a lengthy critique, which she sent to Dr. Axelrod, with copies to the head of the National Institute of Environmental Health Sciences, the Secretary of Health, Education and Welfare, and her congressman and state legislators (Gibbs 8/20/79). The commissioner's report and Gibbs's comments are too detailed to describe fully here, but, among other things, Gibbs attacked the continued lack of any true control group for the DOH epidemiological survey, pointing out that using the wet-dry comparisons or comparisons with residents of streets contiguous to the arbitrarily defined Love Canal neighborhood meant that the controls were also exposed populations.

She again attacked the use of control groups from the literature as the norm for miscarriages at Love Canal, pointing out that the miscarriages at Love Canal were physician-verified, while the control-group figures were self-reported (Warburton and Fraser 1964). She also attacked an experiment described by Dr. Axelrod in which twenty-two pregnant rats placed in a contaminated home had shown no abnormalities, pointing out that the rats should have lived in the home for a period of time before conception took place, so that both females and males would be more likely to show effects. She also questioned the use of such a small number of rats.

Although it was obvious that she had worked with her scientific advisers on the letter, she fully understood the issues, had suggested many if not most of the criticisms, and was able to express them in language understandable to the lay person.

At last the day of the meeting arrived, when the Health Commissioner officially and publicly announced the health-report results to the residents and to all other interested people (Shribman 8/21/79; Porter 8/22/79). The formal gathering of August 21, 1979, took place in a large, echoing room at the neighborhood community center. At the front of the room, seated at a long table, were the commissioners of health and transportation for New York State and other state task force representatives. In the audience were the mayor, the city manager, and other government officials—a total of twenty-one nonresidents, in addition to the news reporters. About sixty residents were present—parents of chronically ill children, women who wanted to become pregnant and were afraid to do so, people fearful of the loss of their lifetime earnings, which had been invested in now-worthless homes, people who had worked diligently for a year in the citizens organizations, and many other worried residents.

Within an hour after the meeting was called to order, the scene again became one in which people tried to convey their fears and anger to their public servants by crying, shouting, and screaming. The officials sat impassively, stony-faced, bravely "toughing it out." The people were crying, shouting, and screaming for good reason, however. Commissioner Axelrod had announced that dioxin had been found in a sample of Love Canal soil, at levels of 5.3 parts per *billion,* and that, in a chemical holding tank at the canal site, it was present at levels of 176 parts per billion. "These findings come as no surprise and further confirm our earlier assumption that dioxin was present in the canal," he was quoted as saying to the audience. The reporter quoting him noted: "Dioxin levels in far smaller concentrations are known to produce cancerous tumors in laboratory animals" (Shribman 8/21/79; see also Carroll 8/21/79).

Dr. Axelrod explained that the DOH planned to warn the construction-project workers about the dangers as well as telling the homeowners and renters about the new findings. He said reassuring things about the dioxin having been confirmed *only* at the southern end of the canal, where the remedial construction had been completed. The words were no balm, for the ground at the southern end had been torn open for repairs to the tiles since late July and would remain open until early in October. Nor was this information comforting to such people as Debbie Cerrillo, who learned that the material was found in both the front and back yards of the home her family had occupied for eight years. The residents also remembered, even though there was no mention of it at this time, the months during which trucks had moved freely from the southern portion of the construction site and over the streets of Niagara Falls, until the practice was stopped in response to the residents' picketing protest the previous winter.

Upon hearing his reassuring words, the mother of an asthmatic child raised her hand. Her eight-year-old daughter had been on long-term visits with various relatives and friends for several weeks, because she suffered acute asthma episodes when she was in her home, located two houses away from the homes that had been purchased the previous year. Trembling, and obviously trying to control herself, the woman asked whether the confirmation of high concentrations of dioxin had influenced the commissioner's recommendations about moving residents. When Dr. Axelrod replied that it had not changed his mind, she sprang from her seat, rushed to the table at which he was seated, and demanded: "Look me in the face! Are you going to tell me I've got to stay another night with my three children in that house in the light of what is being said here today?" With that, this normally respectful woman shredded the card bearing his name and his title and threw the pieces down, crying out, "You are not a doctor! Every time I think of you, I'll think of you with disgust!"

Then a young woman told the commissioner she wanted to have another child but was afraid to conceive while living in the Love Canal area. Dr. Axelrod's reply came in the form of a prepared statement (sent as a letter to Commissioner Hennessy one week earlier):

> Women who became pregnant prior to February 8 obviously were unaware of the increased risks of miscarriage, birth defects, and low birth weights outlined in the Department's study of wet and dry areas of the Love Canal issued in connection with the Commissioner's Order [of February 8]. These women were placed at involuntary risk because they lacked the knowledge of the relevant health data when they became pregnant. Since the availability of the information issued on February 8th, those women wishing to become pregnant are now in a position to make a choice with full information concerning the risks associated with their pregnancy prior to their pregnancy. We have reviewed the requests of the Homeowners Association to relocate women contemplating pregnancy prior to their conception, and

> can find no fair and equitable reason for distinguishing between those contemplating pregnancy and other women in the canal area. (Axelrod 8/13/79).

In short, the women had now been enlightened by an objective evaluation of their situation. They now knew the risks of becoming pregnant while living in the Love Canal neighborhood; their subsequent behavior was their choice and their responsibility, not the government's. The risk, the commissioner admitted in answer to a barrage of questions, was two or three times greater than the risks present in a normal population, and he responded affirmatively when asked whether this meant that a pregnant Love Canal woman had a 30 to 45 percent chance of miscarrying or of bearing a defective child. "What does he mean, 'risk assessment'?" commented one resident privately, "It's more like human sacrifice."

People asked other questions and expressed their negative feelings passionately in the course of the meeting, which went on for over three hours. Finally, people began to drift out. Anxious to get an important question in on another topic before everyone left, Lois Gibbs asked whether some relief action was possible in view of what she called high rates of illness among Love Canal children and some adults as well. She was referring to the preceding three-week period, when many people had telephoned the association office, complaining that headaches, difficulties in breathing, and burning, itching eyes, seemed to be occurring more frequently as the remedial construction work went on. Two months previously, the association had tried to stop the work by an injunction. At that time, the judge allowed the work to proceed, in view of a plan submitted by the task force that included provisions for moving residents out for limited periods of time if a physician confirmed that they were suffering from illnesses related to the construction work at the canal site.

Gibbs had told complaining callers to report to the task force office, had summitted lists of complaints to the health department and had compared the number of Love Canal children absent from the day-care center with absences of children from other neighborhoods. Commissioner Axelrod responded to her question that, since she had simply complained about Love Canal children without submitting adequate control-group findings, her data were incomplete and therefore unconvincing. With that, the entire task force contingent arose, in a hurry to catch their airplane for the flight to Albany. When Lois Gibbs shouted, "What about the rest of the epidemiological report?" Dr. Axelrod replied that she could come to Albany and read it there.

So ended what turned out to be the final report in the last public meeting between the New York State Commissioner of Health and the victims of Love Canal. After the meeting of August 21 and the many events that preceded it, there was no way to return to the hope and trust with which the Love Canal residents had welcomed the health department officials when they first arrived on the scene sixteen months earlier. If anything, the chasm between the groups grew deeper and more intractable over the next year.

Contributions to Knowledge

Lois Gibbs accepted the commissioner's invitation, and she traveled to Albany in September 1979 to learn more about the epidemiological studies. Once there, she was advised that the raw data had not yet been analyzed or described in any comprehensible form.

In fact, although sporadic press releases and the commissioner's order of February 8, 1979, included assertions or conclusions about health, the health-data basis for the statements have been unavailable, with few exceptions. Figures on miscarriages and birth defects among the 97 residents living in the first ring of homes next to the canal were published in the brochure *Love Canal: Public Health Time Bomb* in September 1978. Some data were presented by DOH researchers in closed meetings with unnamed blue-ribbon panels. Some preliminary data describing values for liver-function tests for 27 subjects and sketchy figures on normal and abnormal functioning for 212 otherwise undescribed persons—of the 3,919 whose blood was tested (NYDOH 6/23/80)—are included in a set of documents on Love Canal that was collected by the DOH in the spring of 1979 (NYDOH, 1979, Sec. 57). Even a panel appointed by Governor Carey in the spring of 1980 to assess the health studies done at Love Canal saw very little in the way of DOH data or reports.

Although there may have been some other limited, private disclosures of the epidemiological information and results of blood tests collected in the summer and fall of 1978, they do not change the fact that the scientific norms of publication, open sharing of scientific data, and willingness to describe the basis for conclusions have not been adhered to at Love Canal by the New York Department of Health.

Not only scientific norms were violated at Love Canal; so were standards of respect and obligation of a government for its people. The DOH provided no full reports to the residents who cooperated in the massive health survey and blood sampling that started in the summer and fall of 1978. The residents and their consultants have requested individual and group data to no avail. Commissioner Axelrod's promises to provide "white papers"— that is, data-based explanations for policy decision—have not been kept (Paigen 1/8/79). At one point, when they pressed for data, Dr. Paigen and Stephen Lester were told they could work at the

DOH computer themselves and extract information. When Lester arrived in Albany, however, he was refused access to the health data (Lester 10/25/79).

During the spring and summer of 1979, over 300 residents provided notarized release forms, requesting that Dr. Paigen be allowed to examine the information about them held by the health department, but to no avail. The requests were turned over to the DOH office of counsel (Axelrod 10/16/79), and there they evidently remain.

When he was invited to testify before a congressional subcommittee in March 1979, Dr. Axelrod stated that Love Canal health data exist, but none except cancer registry data by counties appear in the reports of those hearings, even though he apparently agreed to insert Love Canal data in the record (USHR 3/21/79, pp. 290, 295; Eckhardt 8/10/79).

After lawsuits were filed by the U.S. Department of Justice and by the attorney general for the state of New York (in December 1979 and April 1980, respectively), Commissioner Axelrod refused requests for health data from the government lawyers preparing the suits. He contended that the confidential data were protected from revelation by law (N.Y. Public Health Law, Sec. 206, 1 (j)). Both state and federal attorneys have requested the data, so the question of whether data must be provided, and how they will be provided, while protecting the confidentiality of the respondents, will be a matter for the court to decide (see Penca 6/20/80). It is not easy to understand, however, why the data were inaccessible prior to the lawsuits, or why they were not available as grouped data with identifying information removed, for that is the usual practice in the presentation of epidemiological information.

The commissioner's repeated refusals to share information upon request are particularly puzzling, because on other occasions he was willing to share the data. In November 1979, Dr. Axelrod wrote to the president of the National Academy of Sciences, discussing the possibility of a study that never came to fruition:

> I believe that the National Academy of Sciences is the appropriate body to undertake an evaluation of current methodology for the establishment of linkages and the relevancy of epidemiologic approaches to public policy for environmental health. *We are prepared to make available to the National Academy all existing envrionmental [sic] and health data on the Love Canal* to permit this national disaster to be used as a benchmark to be dissected and criticized as a first step in its deliberations (Axelrod 11/9/79; emphasis added).

For a time, there was a rumor that Dr. Vianna was "sitting on the data" while he wrote about it, so that he could be the first to publish. On June 23, 1980, the DOH announced that a manuscript on the DOH epidemiological studies, "Adverse Pregnancy Outcomes in the Love Canal Area," by Dr. Nicholas Vianna and five other scientists (Vianna et al. 1980), had been rejected by the journal *Science.* The document, and the negative peer reviews, were released to the public at the request of the Gannett News Service, under the state's Freedom of Information Law. (NYDOH 6/23/80b). The DOH had argued that the document should not be released publicly because the state was pressing a lawsuit against the Hooker Chemical and Plastics Corporation, but they evidently would have been willing to have the material published in the pages of *Science.* Ironically, not even members of the blue-ribbon panels received copies of the manuscript until shortly before it was made public through the newspapers (Page and Shribman 6/26/80). The document offered little beyond the information on excess miscarriages apparent by the summer of 1978, except for a startling disclosure that miscarriage rates among Love Canal's pregnant women had soared to 50 percent during the mid-1960s.

There the matters stand. As of the autumn of 1981, after three years of effort and the expenditure of $3.29 million (Tarleton 10/7/80), not one study based on health questionnaires had been published in full by the DOH to increase the store of scientific knowledge about the effects of chemicals on the human body beyond what was known by the end of 1978. In April 1981, the New York Department of Health published *Love Canal: A Special Report to the Governor and Legislature.* There are no new data in this document beyond those described in this section (NYDOH, 1981). In June 1981, an article appeared in *Science* (Janerich et al. 6/19/81). This study of cancer incidence used census-tract data for the period from 1955 to 1977. It was not an analysis using the questionnaires filled out by Love Canal residents. The findings are summarized in chapter 8.

The Love Canal residents suffered a deepening alienation during the first Love Canal year. Goaded by their contacts with the DOH and other officials, the residents acted and even thought and felt in ways foreign to them before Love Canal became the poisonous focus of their lives. In the course of that period at Love Canal, the state public-health officials became alienated from the very people whom they were, after all, supposed to serve and protect. By the end of that time, they seemed to be alienated from the scientific values they continued to espouse and, finally, even from basic humanitarian values.

Notes

1. The figures are from four homes, near each other on 97th and 99th Streets. On September 11, 1978, a woman living in one of the houses called the Homeowners Association office, desperately seeking an interpretation

of her numbers, which she could not secure from on-site state representatives. This example is typical of the information given to residents on the quality of air in their homes, usually in their basements. Each resident was provided figures for his or her dwelling only, but neighbors compared their results, as did people in public meetings.

2. Still later, some DOH scientists were to refer to the "adverse pregnancy outcomes" as "biological endpoints" in an attempt to deduce routes of chemical migration (Vianna et al. 1980). See Klaus and Kennell (1976) for a discussion of the intense mourning and grieving by the parents of stillborn babies and neonates who die: "When you feel life, it is a person in your mind" (p. 210).

Merton's (1973) brief comments on a "sadistic type of social structure" seem appropriate in this regard as well. Warning social scientists about the use of important terms, such as "social stratification," he says that a sadistic type of social structure "refers to the kind of conceptual apparatus that, once adopted, requires us to ignore such intense human experiences as pain, suffering, humiliation, and so on. . . . [A]nalytically useful as these concepts are for certain problems, they also serve to exclude from the attention of the social scientist the intense feelings of pain and suffering that are the experience of some people caught up in given patterns of social life. By screening out these profoundly human experiences, the impersonal concepts become sociological euphemisms" (Merton 1973, p. 131).

3. A confidence level of .05 in this case would mean that the probability (*p*) of occurrence of excess numbers of cases of "fetal wastage," low birth weights, and birth defects in the Love Canal population sample, over the number to be expected in a sample of a population unexposed to chemicals had to be so great that based on a statistical theory of sampling error, the numbers indicating the excess would occur by sheer chance, no more than five times in a hundred. Whether their data indicated that *p* level at that time is not known.

4. I have been describing Type I and Type II errors in rather practical terms. Type I and Type II errors are statistical terms describing choices in situations where the probabilities of events occurring are known. Committing a Type I error means a researcher *accepts as true* findings that turn out not to be true. A Type II error occurs with the *refusal to accept as true* findings that turn out to be true. The possibility of one type of error increases as the possibility of the other decreases.

In the Love Canal situation, a Type I error would ensue if it had turned out that the high miscarriage rates determined in the summer of 1978 turned out to be incorrect or unrelated to the chemicals. A Type II error would follow the insistence that the findings do not show anything wrong with the people's health, and then, at a later time, it becomes clear that there really was something wrong. For the decision makers, making a Type I error would mean that resources might be committed *now* unnecessarily, while taking the risk of a Type II error would mean that the discovery of an incorrect decision would be deferred, perhaps for many years in the case of slowly revealed consequences (see Blalock 1960).

6 Problem Solution: Behead the Messenger

Scientific studies and reviews of scientific studies played an important part at Love Canal. It was not the part that the public expects of science—providing definitive facts to resolve uncertainties and ambiguities in a situation requiring protective, collective actions. Rather, the studies and reviews became part of an emerging interpretation of what happened to and what was done for the people of Love Canal. Nelkin (1979), summarizing her review of controversy involving technical decisions, described what was to happen at Love Canal very well:

> Whatever political values motivate controversy, the debates usually focus on technical questions. The . . . controversies develop out of concern with the quality of life in a community, but the debates revolve around technical questions. . . . This is tactically effective, for in all disputes broad areas of uncertainty are open to conflicting scientific interpretation. Decisions are often made in a context of limited knowledge about potential social or environmental impacts, and there is seldom conclusive evidence to reach definitive resolutions. Thus power hinges on the ability to manipulate knowledge, to challenge the evidence presented to support particular policies, and technical expertise becomes a resource exploited by all parties to justify their political and economic views. In the process, political values and scientific facts become difficult to distinguish. (Nelkin 1979, p. 16)

We will start with Beverly Paigen's story.

Stepping out of Line

For the first thirteen years after receiving the doctorate in biology, Beverly Paigen worked as a cancer research scientist, much of that time as an environmentalist, with a special interest in genetic susceptibility to environmental toxins. She had received the doctorate in 1967, after four years of graduate study at the Roswell Park Memorial Institute, a prestigious cancer research center in Buffalo. Once private, the institute is now supported by the state of New York and administered by the DOH. In 1975, Paigen returned to the institute as a senior cancer research scientist, and by 1977 the institute's director, Dr. Gerald Murphy asked her to submit monthly reports on her many research projects, because he was very interested in the rapidly developing field of environmental carcinogenesis.

As a basic researcher, Paigen receives a salary through the institute and is allotted space and some laboratory facilities. Between 1975 and 1979 she brought in almost one million dollars in grants to support her laboratory. Paigen's record in the national scientific community is that of a capable, "solid citizen," actively publishing numerous scientific articles, presenting professional lectures, serving on the U.S. Environmental Protection Agency (EPA) Toxic Substance Advisory Committee and on other advisory boards for EPA, helping to found a thriving environmental program for the State University of New York at Buffalo, and serving as an expert witness on various occasions (Paigen 3/7/80; Sect. 1).

Among other activities, Paigen frequently speaks to and consults with community groups. Paigen believes scientists should work with lay people for the purpose of science, she thinks, is to help people to live better and healthier lives. As a consequence, Paigen sees no role conflict in working with lay people as well as with her scientific peers; she sees such activity as fundamental and appropriate behavior for scientists, particularly those interested in environmental problems.

When Paigen first tested some Love Canal soil, she found evidence of mutagenic chemicals right in the school playground (MacClennan 8/3/78). By the time journalists spoke with her in the summer of 1978, she had already introduced herself to Dr. Glenn Haughie, Deputy Commissioner of Health, offering her help and telling him that she was conducting tests on Love Canal soil (Haughie 7/20/78). Soon after, in answer to reporters' questions, she stated that, in her opinion, all families living adjacent to the canal site should be moved quickly at the state's expense. Although she did not elaborate to the reporters, she knew that working-class people had few resources to simply pick up and leave, and that the DOH order of August 2, had caused the market value of the area's homes to plummet, making it all but impossible for people to sell them.

A few days later, after Stephen Kim, a DOH scientist, released a list of eighty-one chemicals found in Love Canal soil samples, Paigen, as was her practice, responded carefully to a reporter's request for information about the toxic effects of the listed materials. Within a few days, however, she learned that, as a result of the article "Experts Find Carcinogen at Love Site" (Baker 8/6/78), she was "in hot water" with Dr. Gerald Murphy, director of the institute where she was employed. Although she was on vacation, 900 miles from her office, she was ordered to submit a report on her Love Canal work within twenty-four hours. She learned much later that Dr. Murphy thought it was she who had identified the eighty-one chemicals in her laboratory (Murphy 8/6/78; Paigen 3/7/80, Statement and Sect. 6).

She was soon notified that she was to fulfill a set of requirements applying to her alone of all the scientists at Roswell Park Memorial Institute. She was not only to provide monthly reports about her research but she was also to inform the director in advance of all public meetings she planned to attend and of any speeches she planned to deliver—regardless of whether such meetings or speeches were related to her professional work and regardless of whether they took place during or after working hours. She was also directed to report all contacts with the press. Further, she was to outline any research ideas when first conceived, to ascertain that the possible research would meet the Roswell objectives (Paigen 9/19/79).

At that time, Paigen attributed the director's anger and punitive actions to his sensitivity about the public image of his institution. She was annoyed and uneasy about the unusual treatment, but she complied fully with his requests, in the hope that it "would all blow over soon."

Early in September, however, she was chastised for becoming involved in preparation for a collaborative research project with a commercial sponsor without notifying the appropriate administrative authorities (Pressman 9/7/78). The project, a study of mutagenic pollutants from diesel exhaust, was undertaken by the Calspan Corporation. When a Calspan scientist asked for an interested institute scientist, the institute's associate director referred him to Paigen. Paigen has documented the steps she took to inform the administration about the project as it went along, but months later she learned that someone who said he was speaking for the institute director had called Calspan in September and had told the interested scientists that Paigen did not have permission to engage in their study (Paigen 3/7/80, Sect. 8).

Although she did not learn of that call until the spring of 1979, by October 1978, she already realized that she was involved in something more serious than a sensitive boss disciplining a worker who had stepped out of line. She became aware that her boss's bosses were also angry with her.

When Paigen became interested in Love Canal, she began to think about research problems. Over the next two years, she planned or carried out soil tests, tests on small wild animals, and experiments with laboratory-bred mice; she collected fat samples from Love Canal residents who underwent surgery for other reasons; she studied the growth and development of Love Canal children, including dental and hearing examinations; she arranged for nerve-functioning tests; and she examined a variety of specimens provided for her by Love Canal residents. When Lois Gibbs turned to Paigen in September 1978 with her concerns about the possible relationship between the paths of the underground swales and reports of illnesses, Paigen found the idea intriguing and knew it might be important.

Paigen guided Gibbs and the other volunteers in conducting a survey in as organized and unbiased a way as possible, given research resources that were limited to volunteers armed with telephones. The callers collected information about the numbers of persons in each family, the length of time the families lived at Love Canal, and the health problems experienced by family members. This research project was not at the forefront of Paigen's attention, but, when the findings began to accumulate, she grew concerned about them. With 75 percent of the families who lived east of the canal responding, the information suggested that there were clusters and lines of illnesses following the patterns of underground drainage ditches and long-filled-in ponds. In turn, this finding suggested a migration route for chemicals that was very different from the DOH assumptions that the chemicals were moving evenly through the soil.

Paigen tried to set up a meeting with the DOH researchers immediately to share the findings with them. She thought her material might provide leads that would be important for the DOH's trained epidemiologists to follow, using some of the huge resources they were devoting to the Love Canal studies. She got vague answers to her requests for a meeting and no return telephone calls from people who were always unavailable when she called. Finally, Paigen approached William Hennessy, commissioner of the department of transportation and chairman of the interagency task force at Love Canal. He told her to try calling Health Commissioner Whalen again; when she did, Dr. Whalen answered her call.

Determined to present her findings and suggestions to the health department scientists, she was taken aback when Dr. Whalen asked her pointedly why she was doing anything at all at the Love Canal, since no one had assigned her to that study. Paigen responded that, as a professional researcher, she did not wait for assignments but rather chose the research areas she thought appropriate to her interests and that Love Canal was appropriate to her interest, both in aquatic sediments and in family patterns of genetic susceptibility to environmental toxins. Although she answered politely, and in a forthright fashion, her feeling of unease, aroused by her difficulty in reaching the DOH researchers, was now strengthened. Since Roswell Park Memorial Institute is part of the DOH, Dr. Whalen was the chief of the state agency employing her. Now, she thought, perhaps the commissioner was offended by a newspaper article about the possible correlations between illnesses and the underground drainages at Love Canal (Brown 10/4/78). She recalls that she felt "like I was sitting on a keg of dynamite." Although she sensed that someone powerful was vaguely displeased, she believed that her information was crucial to the DOH research efforts and to future policy decisions about the Love Canal residents. Thus, she persisted in requesting that a meeting be convened between her and the DOH researchers most involved with Love Canal: David Axelrod, Glenn Haughie, Nicholas Vianna, and Stephen Kim.

Dr. Whalen agreed to call the meeting. However, when Paigen made a routine request later in the day to use funds from her research grant in order to travel to Albany for the meeting, she was refused. The institute's fiscal officer explained that "the Commissioner's office" had telephoned and said they would not approve such a request. Nor was her request to count the day in Albany as part of her regular work day honored (Paigen 3/7/80, Sect. 9). She went to Albany at her own expense, but later she realized that the director's refusal was, in effect, a denial that her work at Love Canal was a part of her legitimate research efforts. (Some of her later travel to Albany in regard to Love Canal was funded by the DOH.)

By November 1, when Paigen met with the DOH researchers, the study had already been used politically by the Homeowners Association president, Lois Gibbs, who had released some of the preliminary results to the newspapers. When questioned by reporters, Dr. Vianna of the DOH had firmly rejected the results (Shribman and MacClennan 10/18/78). However, at that November 1 meeting, the DOH researchers expressed interest in the maps and in the house-by-house illness counts that Paigen described. Paigen had assumed that the people she thought of as colleagues would be interested in the serious scientific implications of her small study, and she assumed they would overlook the use made of it by a group of organized lay people.

Paigen left the meeting feeling that it had been worthwhile to persist, for she had accomplished her aim of alerting capable professionals to the research leads indicated by her limited study. If clusters of chemical-related illnesses were to be found along the paths of underground waterways or swales, crosscutting through the old canal, there was an alternative hypothesis to be tested, along with the DOH's chief hypothesis that the chemicals were oozing evenly from the old canal. She now thought the DOH scientists were going to examine their data with both hypotheses in mind.

Two weeks later, however, a news article announced that state officials had "serious reservations" about the theory that toxic chemicals may have migrated along the paths of underground streams. "Dr. Axelrod said," according to the reporters, "the association evidence was not gathered in a scientific manner and said other phases of the state's examination of the Love Canal problem were set back while scientists evaluated the homeowner's theory" (MacClennan and Shribman 11/16/78). An official chronology of events, composed two years later, described the DOH in the midst of a similar but much broader study of underground swales at the time (N.Y. Health Planning Commission 1980).

As Paigen thought about it later, she realized that, by the November 1 meeting, in addition to embarrassing the Roswell director with "unseemly behavior," she had evidently angered the health commissioner by what he perceived as her unauthorized participation in a health problem being addressed by his research laboratories. Furthermore, she realized that the DOH scientists may have felt that there was some collusion between Lois Gibbs and herself to exert pressure on the state to purchase more homes by publicizing the results of the Homeowners' Association survey. The DOH scientists may have felt that, in taking "an uneducated housewife's" ideas seriously, Paigen was deflating the image of scientists as the keepers of esoteric knowledge. There is another possible reason for their public rejection of the Paigen suggestions at that time, however: Paigen may have been interfering with the process of problem definition.

Defining the Problems: Money Again

Paradoxically, the answers we already know may influence what our questions can be. In the autumn of 1979, the New York agencies addressing the Love Canal problem may well have been engaged in the process of deciding what questions would be addressed at Love Canal, with the very definition of the questions dependent on the available solutions. Throughout that period, state officials were trying—in vain, as it turned out—to obtain federal funds for various aspects of the research and the construction project and for the evacuation of people from Love Canal—beyond the $4 million already pledged by EPA for a demonstration project in the area.

As late as November and December of 1978, letters to federal sources of funds included such phrases as "Because of the health threat, the most immediate need is to evacuate the affected residents" (Flacke undated, fall 1978); "Exposure . . . to the entire area . . . constitutes a threat to health" (Grushky 11/24/78); "Preliminary findings have led to the suspicion that the chemicals in certain areas outside rings one, two and even ring three may constitute a threat to health" (Kistler 11/9/78); and "In this situation, there is still no certainty as to the full spread of toxic materials, the total number of persons affected or the adequacy of currently planned emergency measures" (Grushky 11/24/78).

While these urgent letters and applications were being prepared quietly and sent by a variety of state officials, the public stance was that there was no evidence of problems at Love Canal beyond those that could be contained within the area designated in August for remedial construction and for home purchases. Caught up in a sort of holding pattern, the officials might have felt that, if funds became available to seek solutions, they could define the problem as more widespread, but that there was no point in disturbing the populace with further statements about the possibility of serious health hazards without financial resources to move people out. The fiery reaction following the August 2, 1978, health department order, with its recommendations for people to move out but no provisions for implementing such moves, may have taught the necessity for caution.

In such a situation, it may have been more than annoying and embarrassing to have a reputable scientist such as Paigen making firm public statements about health hazards and the state's obligation to move people out of Love Canal. If Paigen's statements were taken seriously, with a public admission that the new hypotheses were worth pursuing, then the demands of the organized citizens would be legitimated—not only by one respected scientist but by the health department itself. Then the chaos of August might be repeated—and worse.

Getting a Contract Out

After the November 1, 1978, meeting, Beverly Paigen learned that her Love Canal interests were going to influence her life and work in ways she had never anticipated. After the meeting, she turned her time and attention to a grant proposal to EPA (unrelated to Love Canal) that had been in process for over a year. The planned work was to be a collective effort among scientists from several New York State agencies and was to be administered through the New York Department of Conservation (DEC). Paigen's contribution was important, because she had conducted the pilot study on which the proposal was based, she was particularly adept at one of the proposed research techniques, and she had formulated major ideas for the research proposal. Participants who were DOH employees required the health commissioner's approval to participate—normally a routine matter.

By November 20, 1978, on the basis of a very rough draft, Paigen's contact person in DEC told her that the general lines she and the others suggested looked very good and that they should move quickly toward preparation of the final draft of the proposal. After that conversation, she telephoned Roswell's financial officer for instructions on routing the proposal through the bureaucracies involved in a multiagency endeavor. He told her that the director wanted to know what she was planning to do and insisted that she submit the rough draft immediately. Aware of the delicate situation she seemed to be in with the institute director, Paigen complied, cautioning him that the proposal draft was in a preliminary, sketchy stage. Two days later, the project's collaborators revised the early drafts thoroughly. By November 27, Paigen completed her portion of the proposal and mailed it to the project coordinator, to mesh with the other participants' materials. On the following day, Paigen was astonished to learn that the institute director had forwarded to the health commissioner the early, rough draft she had sent to him so reluctantly. Paigen wrote to Commissioner Whalen immediately, notifying him that the rough draft had been sent by mistake and telling him to contact the project coordinator in DEC for the final, finished proposal, which was now ready (Paigen 11/28/78).

Dr. Whalen had already communicated with her director, Dr. Murphy, however (Whalen 11/28/78). On December 4, one full week *before* Paigen formally submitted the completed joint proposal to Dr. Murphy for routine administrative review and sign-off, Dr. Murphy wrote to various Roswell officials: "Dr. Paigen's contract is not workable and not one that meets with Dr. Whalen's approval" (Murphy 12/4/78). Paigen had received no notice of the decisions the commissioner of health and the director of Roswell had already made by the time she submitted the proposal. She did not learn of those decisions until much later.

When Dr. David Axelrod assumed his duties as commissioner of health in January 1979, Paigen was certain that her relations with that office would improve. She respected and trusted Axelrod as a man of ability and integrity, and she felt she had a cordial working relationship with him. On one occasion, when they were meeting about Love Canal studies, she mentioned the administrative red tape about the grant proposal and her embarrassment when the rough draft was sent to the commissioner. She told him she did not comprehend the problem, and she recalls that he told her that he had no objection to her participation in the project and that, if the proposal were funded, she could participate.

In April 1979, the EPA awarded the contract to the New York Department of Environmental Conservation for the proposed study, and the work was scheduled to commence on May 1. Paigen soon became seriously concerned, because, although other people on the project began to prepare to work, the routine notice that her subcontract was funded did not arrive. When she traced the matter, she made the disquieting discovery that the delay stemmed from the office of Commissioner Axelrod. By the time she learned the source of the problem, however, the good relations between her and the commissioner were but a memory.

February 8, 1979, Revisited

In mid-January 1979, after Dr. Axelrod became health commissioner, Paigen felt he treated her with more respect than had the previous commissioner, and that he was taking seriously the hypothesis about the correlation between the location of swales and of illnesses. In fact, by late January, Paigen was somewhat chagrined that the DOH researchers were talking about historically wet and historically dry areas at Love Canal, hinting that *they* had first thought of looking at the correlations with illnesses. At that time, relations seemed to be improving so much that she was surprised when she was invited to the first but not to a second meeting conducted by Commissioner Axelrod with a group of outside experts. It could be that the reason she was not invited to the second meeting was related to a local front-page news article following the first meeting, which cited Paigen as calling for the evacuation of more families from the Love Canal area (Brown 1/23/79).

The DOH's second meeting with outside experts was held on February 7. This one was called a blue-ribbon panel meeting, with expert epidemiologists, toxicologists, and pediatricians consulting with the DOH scientists on the results of the DOH survey and clinical reports. No consultants to the Love Canal residents were invited to this blue-ribbon panel meeting, nor were the names of the experts revealed by the DOH at that time. Paigen felt that the residents were not treated with respect when she, their consultant, was excluded from this important meeting. Moreover, the idea that the panel members' names were kept secret was an affront to her sense that scientific decisions underlying policy decisions should be made openly and that the public has a right to know who makes decisions and on what basis.

On the night of February 8, 1979, an important meeting (first described in chapter 5) took place between the state officials and the Love Canal residents. The residents were hoping that a decision would be forthcoming to get them all out of the situation—either by moving them or by declaring that the residents' homes were safe. Many state officials, many residents, some policemen, and many members of the press were present at the meeting.

When Dr. Axelrod addressed the audience, he announced that he was extending the health order of August 2, 1978. He said that the DOH had had the Homeowners Association's full cooperation, and that the DOH's use of the leads they had provided in the Paigen survey showed that government and citizens could work together in cooperative efforts. Then he announced the newest policy. If they requested it, families with pregnant women and children under the age of two living on the streets to the east of the inner rings would be moved to temporary living quarters until the youngest child reached age two. Even though Axelrod emphasized that moving the children under age two was a sign of the extra precautions he was exercising, his announcement was not received with gratitude by the people. They did not agree that his recommendations represented extra precautions to guard their health. In fact, the first responses to his speech were combinations of questions, shouts, and curses.

The governor's aide, Thomas Frey, valiantly asked everyone to be quiet and polite, but the more Dr. Axelrod calmly stated that there was "no evidence" that children in the area had been exposed to risks greater than those in other areas, the more upset people became.

What was happening in this discourse between health commissioner and citizens? The commissioner had convened a high-level panel of consultants, had made recommendations, and was conveying policy decisions to the residents. The residents were repudiating him and all that they thought he stood for. Not only did they refuse to treat him with respect for himself or his office, they also told him that they doubted his department's competence, his professional procedures, his conclusions, and, furthermore, his integrity and the integrity of the department and of the state government. As far as they were concerned, they were receiving mixed messages from him—that they were *not* safe but that he would like them to *think* that they were; that he was *not* taking extra precautions but that he would like them to *think* he was doing so.

In the midst of the vilification directed toward Dr. Axelrod, Dr. Paigen asked to speak. The audience calmed down immediately, paying her strict attention. There was a striking contrast in the audience's rapport with her and their attitude toward the commissioner. When Paigen spoke, she used a neutral tone, expressed respect for the DOH scientists, and cushioned her remarks in statements to the effect that scientists interpret information differently, that scientists do disagree, and that that is the way of scientific work. However, what she said could have been interpreted by people in positions of responsibility—those who, at the moment, were literally on stage—as highly critical and obviously influential comments, lending credence to the angry crowd's interpretations.

Paigen calmly told the now-quiet audience that the state had examined data for miscarriages and for fetal deaths and had decided to move the people who were most at risk—those with fetuses or very young children. She said the state had not yet examined the rest of the data—a contradiction to Dr. Axelrod's statement that there was "no evidence" that Love Canal children had been exposed to risks greater than those in other places. She pointed out that the information the members of the community had volunteered to the Homeowners Association callers had shown there were other illnesses—again a direct contradiction to Dr. Axelrod's statement and implications. She said that, in her opinion, once the information was collected and analyzed by the state DOH, it would justify the removal of additional families—a position Axelrod had studiously avoided taking. She also stated that, in her opinion, the central nervous system effects were the most compelling data—something Axelrod had not even mentioned—and that central nervous system effects might be more compelling than fetal damage—casting doubt on the indicators Axelrod chose to emphasize. Finally, she said that she hoped that the state would see the facts once they examined the data and that they would then make further recommendations.

Paigen thought of herself as a scientist and, thus, a free-thinker. It is possible that some DOH people and others considered her a heretic, a traitor who was supporting "the other side."

Going Public: Send in the Men, Save the Canaries

Until the week of the February 8 meeting, Beverly Paigen had had a good deal of trust in Dr. Axelrod and his judgments. She was shocked, however, at his limited recommendations that the state underwrite temporary moves only for families with pregnant women or children under two. There was an important difference of opinion about the meaning of fetal indicators. Dr. Axelrod stated that the fetus was the most sensitive indicator of chemical-related distress and that, by including children under age two in the order, he was taking extra precautions to protect people's health. Paigen reasoned, however, that, if the fetus is the most sensitive indicator, everyone should move out of the area immediately. She drew the colorful analogy that, in a previous era, when canaries were sent into the coal mines to detect the presence of gas, the death of the small birds did not lead to decisions not to send more *canaries* into the mines while continuing to send in men who were bigger and stronger than canaries. In short, she argued that the fetus as indicator was being used incorrectly if the policy implication was to remove the fetuses.

Paigen was concerned that, with the new possibility of getting out, more health damage would ensue if couples decided to use pregnancy to solve their problems—"ejaculate to evacuate," as the earthy slogan went. She felt obliged, furthermore, to report on the telephone survey in which so many residents had participated. Twelve days after Dr. Axelrod's meeting with the Love Canal residents to explain his most recent order, Paigen spoke to a public meeting of about one hundred residents and for the first time publicly explained the full results of the Paigen–Homeowners Association survey. She advised the audience not to contemplate pregnancy until several months after leaving the Love Canal area in order to allow the chemical matter stored in the fatty tissues to be slowly excreted from the body well before conception took place. The greatest danger to fetal life, she pointed out, was during the first days and weeks after conception. If they remained in their Love Canal homes, by the time a pregnancy could be confirmed, applications filled out, requests granted, and new housing located, damage might already be done to an unborn child.

Illustrating her lecture with slides and simple tables, she showed why she thought that not only were there strong indications of greatly increased rates of miscarriages, stillbirths, and birth defects among residents of wet as compared to dry areas, but there were also indications of higher rates of nervous and urinary-system disorders. When she started her presentation, several television cameras and microphones were directed at her. She calmly stated that, in her opinion, many more people, at least another 236 families, should be moved at least temporarily from the wet areas and that the criteria should simply be the location of their homes. She stated also that, although pregnant women and children under age two should be evacuated from the wet areas immediately, families living in dry areas should also be moved if they so desired (also see MacClennan 2/20/79; Dearing 2/21/79).

The Love Canal audience sat quietly throughout her presentation and then asked some practical, technical questions about the findings and their implications. Paigen explained probability values to them and was thanked by an audience member for clarifying that mystery. She stated that, in her opinion, additional laboratory studies should be done instead of relying almost solely on epidemiological surveys to assess the effects on the residents' health. The audience listened intently. Most left the meeting as soon as it was over, certainly not happy with the news she had brought them but satisfied that they had been told the results of the study in which they had participated. Some people stayed and spoke with Paigen further about the implications of the findings.

This well-publicized meeting simply underscored and made more formal the disagreement Dr. Paigen had expressed, however politely, with the commissioner's interpretations and recommendations at the meeting the previous week. There were, of course, vast differences in the positions occupied by the commissioner of health and by Paigen. Because she had the luxury of not occupying a decision-making position, she could express her opinions without having to find the resources to act on them. Dr. Axelrod claimed, however, that he did not make the decisions but simply made recommendations. This point was frequently raised in the meetings between residents and members of the DOH. Residents would ask: "If you don't make decisions, then why won't you at least *recommend* that we all be moved? Why won't you support us?" There never was an answer to this.

Going Public: Testifying under Oath

Paigen continued to speak out publicly. In March, 1979, called to testify before a congressional subcommittee investigating the problems of hazardous-waste disposal, she presented a detailed report based on the Paigen–Homeowners Association telephone survey research, since she did not yet have results from other studies, which were still in early stages of development. Her report showed higher rates of reported illnesses in wet as compared with dry home locations at Love Canal. That testimony became the subject of an attack on her over a year later.

In her prepared testimony, Paigen presented the information that led her to suspect that toxic chemicals were migrating along the paths of the old streambeds (swales) and that the people living in homes near these paths might have increased rates of miscarriages, birth defects, nervous breakdowns, and diseases of the urinary system as compared to residents in dry areas. She stated that she thought at least 140 and up to 500 families should be evacuated from the Love Canal neighborhood, in addition to those already moved. She described the history of the study and carefully described the study's limitations:

> This survey suffers from several problems. First, a lay person reported diseases to a lay person and some of the people involved may not understand the true nature of their illnesses. Second, both the people reporting and the people collecting the information have a vested interest in the outcome and there may be over-reporting of disease. And third, I did not have any resources so I could not verify independently the reports of disease with physician records. To overcome these problems I concentrated primarily on those health effects that are diagnosed by a physician and that a lay person knows by name. To correct for over-reporting I used internal controls in the neighborhood [that is, no one reporting knew whether they lived in wet or dry areas, or that these factors were even being considered]. (USHR 3/21/79, p. 61).

She pointed out that the health effects she presented might be serious underestimates of the true health effects. First, she did not have a true control population, for she was comparing people from possibly *more* exposed homes to those in *less* exposed homes, not comparing them to people in unexposed homes. Second, she had not studied the 239 presumably most exposed inner ring families. Third, there was a possibility that some people with serious health problems would not cooperate in a study conducted by neighbors.

Using aerial photographs and street maps, Paigen showed the location of the old underground drainages and the location of all the homes included in the survey. More than 75 percent of the families living east of the canal participated in the survey and those who did not participate lived in homes scattered haphazardly through the neighborhood. Her presentation of findings included simple tables, statistics she had prepared with a statistical consultant, and maps showing the location of the homes with the illnesses she was describing.

Dr. Axelrod testified before the same congressional subcommittee the following day. His written testimony was not focused on specific health effects but rather provided a general history of the DOH's work at Love Canal and a statement on the lack of cooperation the state had received from the federal government in funding and guidelines, despite the huge problem of toxic wastes (USHR 3/21/79, pp. 287-313).

During the question-and-answer portion of the testimony, Dr. Axelrod testified that, although he did not agree with Paigen's data on the *level* of risk, he did agree with her conclusions about pregnant women being at greater risk if they lived at Love Canal.

When questioned by Congressman Albert Gore (D. Tenn.) about the lack of control groups for the DOH study, Commissioner Axelrod explained that they were not used because the EPA had refused to fund them, at $1.5 million for each of several control groups (USHR 3/2/79, p. 291).

The commissioner was asked how he could justify protecting only the most sensitive indicators, the fetus, "and not doing anything with regard to the remainder of the people." He responded: "We have not been able to substantiate or find evidence that would indicate to us that there has been an adverse response for other members living in that area, particularly the adults" (USHR 3/2/79, p. 293).

It is instructive to compare the question-and-answer sections of the testimonies of Dr. Paigen and Dr. Axelrod, for many of the questions asked of Axelrod seemed to stem from Paigen's oral testimony on the previous day. At one point, for example, Paigen had stated:

> [W]hen a governmental agency, particularly scientists in a governmental agency are doing studies like this . . . when they make a public statement or a decision, they should release a white paper that gives the scientific methods they used, the questions they asked, how they did their study, what the results are, what the statistical analysis is? Instead, all they do is make claims in press releases, and we have no way for independent scientists to review the validity of their conclusions. (USHR 3/21/79, p. 83).

The following day, the congressmen questioned Dr. Axelrod at length about the confidentiality of the DOH data and about the secrecy surrounding the names of the blue-ribbon panel members. Dr. Axelrod cited a provision of the New York Public Health Law (Sect. 206(1)j) and also stated that there had been a threat to kidnap someone on the task force—implying that, therefore, the blue-ribbon panel members must be protected (*Niagara Gazette* 3/23/79*a*). Congressman Bob Eckhardt of Texas responded that he was puzzled and disturbed that a congressional subcommittee, as well as the people affected, could not learn the names of the panel members (USHR, 3/21/79, pp. 307-308). The congressman asked Dr. Axelrod to submit the state's complete study with the data that had been collected. Dr. Axelrod explained that large portions could be submitted but that "a certain section of the State's data, like discussion with expert panels was conducted under a confidentiality provision" of the public health law (USHR 3/21/79, p. 290). If such data were submitted, they do not appear in the publicly available documents of the hearings. Only the rates for malignant neoplasms in New York State for 1976 to 1978, including a county-by-county breakdown, are included.

More than two years later, the names of expert committee members were included in the New York Department of Health's *Love Canal: A Special Report to the Governor and Legislature* (April 1981, p. 32). However, the health commissioner's refusal to divulge the names of expert consultants at the time policy decisions were made had earlier distressed the residents. Dr. Axelrod's refusal to provide names to the Eckhardt committee was supported at the time by the DOH counsel (USHR 3/22/79, p. 311).

Two months after the House subcommittee hearings, in May 1979, Paigen repeated her testimony before a Senate subcommittee investigating the same problem, hazardous-waste disposal, with Senator Daniel Patrick Moynihan of New York presiding (U.S. Senate 5/18/79). Once again, in another prestigious forum, Paigen's findings, conclusions, interpretations, and policy recommendations were taken seriously. Now Paigen's frank criticisms of the DOH and of its commissioner, in the form of sworn statements, were matters of public record.

Meeting with the Governor and Other Audiences

In late April 1979, Governor Carey consented to meet with Paigen and the Homeowners Association president and vice-president, Lois Gibbs and Debbie Cerrillo, in a closed-door session during his visit to the region for other purposes. Commissioner Hennessy and DOH Commissioner Axelrod were present at this April 26, 1979, meeting, as well as the governor's aide, Jeffrey Sachs. At the meeting, Dr. Axelrod pointed out to Paigen, Cerrillo, and Gibbs that Governor Carey had been very generous with the Love Canal people and had spent more than anyone else ever had for recovery from a man-made disaster, that this Governor had stepped in and had demonstrated his compassion for the Love Canal people.

The governor, however, was reported to be rather abrupt and gruff in his manner with the Love Canal representatives and even more so with Paigen. He questioned her work and her lack of medical credentials. She countered with her extensive research knowledge and her doctorate, the highest scholarly degree attainable. She handed him a copy of the report she had prepared for her congressional testimony and started to present some of the highlights to him. She began telling him that, even though the data were self-reported, people had had no way of knowing whether they lived in wet or dry areas when the study was done. Governor Carey interrupted to ask whether her report included verification from physicians for all the reported illnesses. In trying to explain that the data included only the illnesses that people stated had been treated by doctors or hospitals, and that the DOH had the data to confirm these reports, she started to say, "No, but . . . ," at which point the governor snappped that her report was useless and threw the twenty-page work at her. As Paigen described the next few moments:

> The pages fluttered all around and landed everywhere. I slowly picked up every sheet of paper from the floor and table, put them all in order, shuffled the pile to straighten them out—it was real quiet in the room the whole time—and then I just said, "Now I'd like to finish telling you about my findings on the people's health at Love Canal." And I did. (Paigen 5/79)

The governor who had calmed down by then, listened and then promised that a report would be forthcoming explaining the health evidence that had led to the February 8 decision to move temporarily only families with pregnant women or children under aged two. At this writing, over two years later, the promised "white paper" has not been produced.

Meanwhile, Paigen continued to spread her message to ever-wider audiences and to seek interest and support from the scientific as well as the lay community. She participated in many local events, such as lectures and workshops, including a scholarly forum at the State University of New York at Buffalo that was open to the public. There she stated, before television cameras, that there was a conflict of interest for state scientists to be engaged in health research at Love Canal, because the state would have to foot the bill for adverse findings (*Niagara Gazette* 3/23/79*b*).

An Evaluative Review by HEW-EPA Scientists

After Commissioner Axelrod's recommendations of February 8, Congressman LaFalce acted again. Paigen had spoken to him previously of her concern about the way the DOH was "sitting on its data." Through his intervention (LaFalce 2/22/79) HEW Secretary Joseph A. Califano impaneled a group of scientists from HEW and from EPA, headed by Dr. David Rall, chief of the National Institute of Environmental Health Science (NIEHS). The group's task was to examine the differing data related to the Love Canal health studies and the differing interpretations made by Dr. Paigen and by the DOH scientists. In preparing to meet with the scientific panel, Paigen wrote a memorandum to Rall, listing her criticisms of the DOH studies and expressing her concern that the health hazards at Love Canal were not accurately assessed by the DOH. She asked: "Please do not view [the criticisms] as a running battle between me and the State—a battle that I have no desire to be in" (Paigen 3/29/79). She also sent a document she had prepared and distributed previously (Paigen 12/19/78), criticizing the health department's analysis of miscarriage rates at Love Canal. Among other things, she questioned their reliance on a comparison group drawn from the literature (Warburton and Fraser 1964) as the standard for assessing the miscarriage rates of Love Canal women. Even though the Love Canal rates were higher than those of the comparison group described in the article by Warburton and Fraser (1964), Paigen thought the rates of excess miscarriages were actually underestimated for Love Canal women. The comparison group were women interviewed at a hospital, who had previously borne a defective child. Most important, all their self-reported miscarriages were counted, while in the Love Canal group only physician-or hospital-verified cases were counted. [At a public meeting in July 1980, Dr. Warburton herself replied to a question from the audience, that the group she had studied did not provide an appropriate comparison for the Love Canal women, and that using her study as a control would result in an underestimate of the excess miscarriages at Love Canal.] In her memorandum to Rall, Paigen also specifically criticized the DOH researchers' use of the arbitrary .05 probability level as a value decision; the choice of cut-off dates for current and former residents; the minimizing of current risks by averaging results over long periods of time; the unequal treatment of positive and negative cases (that is, ruling out positives for any other possible causes but not applying the same analysis to ascertain that seemingly negative cases were not, in fact, showing effects); incorrect classifications of homes into wet and dry areas; lack of control populations; the DOH's failure to accompany public statements and decisions with data that could be independently reviewed; and their behavior with the Love Canal people, which had eroded the residents' confidence and might already have resulted in residents not cooperating fully with any state-sponsored study (Paigen 3/29/79).

In July 1979, after reviewing written information and talking with both Axelrod and Paigen, the Rall panel issued a public statement tactfully supporting the idea of a "gradient" of health effects depending on location of homes. They suggested: "Prudent public health practice dictates that these exposures be minimized to the extent feasible"; and they encouraged the continuation of studies at Love Canal. The committee stated that the public might perceive a conflict of interest in the fact that the state employees were conducting health studies; therefore, the use of outside scientists "both in the interpretation of data and formulation of recommendations to the State should be continued. . . . the State may also wish to include non-scientists, local residents and others in future deliberations" (Rall 7/26/79).

Although the language and tone of the report were circumspect, the message seemed clear. Despite the enormous difference in resources, not only were Paigen's study and the DOH study considered at least equal in merit, but Paigen's work was publicly supported (Powell 7/26/79). In addition, HEW Secretary Califano impaneled a group of HEW and EPA scientists, partly as a consequence of this review, to prepare a full report on assessing the potential health risks from chemical dump sites. (USDHEW 7/26/79, 2/80). By this point, without the invitation of the state health commissioner, federal agencies were now publicly reviewing the DOH study of Love Canal problems. Moreover, the federal officials seemed to be "carrying the ball" on the national level before the state scientists had the opportunity to present their work to the scientific community. For all this, Dr. Beverly Paigen could be thanked.

Harsh Consequences

People who unveil views contrary to official corporate or other bureaucratic policies often suffer harsh consequences for their acts. In this case, the opinions Paigen expressed were not simply innocent outcries reinforcing what people suspected to be true. Her statements, behavior, strongly expressed opinions, and use of scientific credentials and collegial networks gave legitimacy to the Love Canal residents who were struggling to get out of their situation and who were using every pressure tactic they could think of in the attempt.

In the summer of 1979, Paigen learned that Health Commissioner Axelrod had refused permission for her to begin working on her part of the large subcontract she had been awarded by EPA. Paigen's protests concerning the commissioner's treatment of her grant were supported by the Council of the Association of Scientists of Roswell Park Memorial Institute. They were more impressed with the excellent peer reviews of the project than the commissioner's comment that the grant was "poorly conceived" (Axelrod 8/15/79). Their letter to the commissioner described his actions as a "disturbing departure from existing procedures" (Chapman 9/1/79).

Dr. Axelrod did not reply to the council's letter, but decided instead to bring the matter of Paigen's treatment by DOH and by the institute administration to the Roswell Park Board of Visitors. Paigen wrote the board detailing what she considered demeaning, time-consuming reporting requirements, different in kind and degree from those placed on other Roswell scientists, and noting that Dr. Axelrod had blocked her participation in the EPA subcontract (Paigen 3/7/80).

The Board met three times. On June 2, 1980, Paigen received a letter from the board's chairperson, essentially supporting the commissioner and the institute administration. It was probably unrealistic for her to expect a different result. The board was not an uninvolved, neutral body. In the conflict between bureaucratic and scientific norms, boards tend to side with administrators rather than with scientists whom they view as employees. Moreover, in the course of their duties, Roswell board members related personally to the health commissioner who is an exofficio member of that body. (The board chairperson explained the health commissioner took no part in making their decision.) Members of the board are also appointed by the governor. One can speculate that the episode illustrates the widely recognized problem of what happens to troublemakers in bureaucracies (Fox 5/30/80).

Paigen experienced other problems as well. More than once, colleagues told her privately that people were nervous about participating with her on Love Canal research without an invitation from DOH (MacClennan 5/25/80). Mail addressed to her at Roswell, coming from EPA, the Environmental Defense Fund, or from the Love Canal Homeowners Association was delivered opened, the envelopes resealed with tape. Postal authorities confirmed that the mail had been delivered in proper condition to the institute. Mysterious accidents happened to some of her specimens. On one occasion, Commissioner Axelrod chided her for using institute stationery and a 15-cent stamp in writing to him about her Love Canal work (Axelrod 10/16/79).

More distressing was her encounter in August 1979 with her first New York State income-tax audit in twenty years. Upon meeting with the auditor, she discovered that her tax folder bulged with news clippings of her Love Canal activities. After Paigen complained, she received an apology from the State Taxation Commissioner. His letter explained why "her fears of bias in the selection of tax audits" were unfounded. Her return, he explained, was number 757 of 1500 cases ranked according to the possibility of "erroneous tax reporting." The commissioner said:

> During the course of this audit, you discovered that your tax file contained 1979 and 1980 newspaper clippings concerning your involvement with the Love Canal Homeowners Association. . . . Since at least 1974, we have asked field people to clip and pass on to managerial staff local news stories involving cases of federal tax evasion and income tax delinquencies or fraud. . . . I wish to offer my apologies for the . . . errors in procedure which have inconvenienced and concerned you. However I hope that you will agree that, in all fundamental respects, the tax audit of your 1976 return followed normal, objective procedures. (Tully 7/11/80 and *Buffalo Evening News,* 7/25/80).

A year later the allegedly 757th most-likely-to-be-erroneous return in a group of 1,500 selected for complexity, was found short by some $600, related to a deduction Paigen's husband had declared for lawyer's fees.

Paigen did not respond to sanctions on her behavior by conforming quietly, but instead resisted all attempts to curb her. As time went on, despite the mounting sanctions, she increased her active support of the Love Canal residents—support beyond the usual boundaries of scientists who study subjects but do not become involved with people. Her lone voice—quietly, clearly, and persistently dissenting from the scientific rationales provided for decisions made about Love Canal problems—continued to underscore the political and value bases of those decisions.

Beheading the Messenger: Year Two

Early in the first Love Canal year—a period extending from the spring of 1978 through August 1979—the Love Canal area was declared a hazard to the health of residents. The first declaration, in June 1978, was based on the presence of toxic chemicals in the air, soil, and sump pumps of the neighborhood. The next declaration, in August 1978, was based, in addition, on preliminary health data showing excessive rates of miscarriages and birth defects for women living adjacent to the canal. Note, however, that the governor's decision to purchase homes, the designation of which homes would be purchased, and the designation of an area for remedial construction were all made well *before* the launching of data collections for environmental and health studies beyond the preliminary ones.

In order to comply with funding regulations, the decision to purchase homes was said to be based on the need to use the backyards of some homes for construction purposes, rather than on the evidence of health hazards for residents of those houses. When Dr. Paigen presented the findings of the Paigen–Homeowners Association survey to the department of health, the study was publicly rejected by the DOH *before* the DOH data that might have supported her findings were examined. In contrast, three months later, when the health commissioner announced that the state would subsidize moves for families with pregnant women or children under age two, the official order was said to be based in part on DOH information supporting variations in possible health effects in wet versus dry residence areas, first publicly reported in the Paigen–Homeowners Association survey. The next day, the offer of a state-supported move was extended to pregnant women and children under age two living in a Love Canal neighborhood area west of the canal, chiefly in the low-income projects, where health data had not yet been examined. That area had a high percentage of minority-group residents, and the delicate political—not solid scientific—reasons for the decision to include them in the new decision were obvious to everyone.

In the third week of August 1979, Commissioner Axelrod announced that he would *not* recommend the move of families contemplating pregnancy, despite the twofold to threefold increase in risks of miscarriage for Love Canal women. The residents had been duly warned of the risks, he said, and thus could decide for themselves whether or not to become pregnant.

The events of the second Love Canal year—late August 1979 through late autumn 1980—involved more actors, and the stage expanded to include the federal government. The social and political processes among government agencies and between Love Canal residents and the agencies, the statements of Hooker Chemical Company, and the interactions of all these with the media bore a striking resemblance in the second Love Canal year to the first-year events described in previous chapters. It is not surprising that, during the second year, scientific studies and reviews of those studies were used in decisions about the Love Canal problem in a manner not very different from the way they were used in the first year. Some studies served as catalysts, forcing powerful people to do something about the fate of the Love Canal residents. Each step, however, seemed to be prompted by reactions to the *public release* of preliminary findings, not by conclusive evidence of increased health hazards for residents in the Love Canal area. It was not that such conclusive evidence existed and was not utilized; rather, the path to collecting and analyzing such evidence was blocked by the behavior of the responsible agencies.

In the second Love Canal year, despite new actors in some of the roles, the unfolding drama was repeated: first, the actors ignored the situation as long as possible, then denied responsibility, then rushed in to do something when preliminary studies implied hazards to people's health, then minimized when the extent of the problem relative to resources became apparent.

Although this pattern of behavior makes for exciting news-as-entertainment, it is not compatible with good scientific research. Research requires reflective thinking, planning, testing, examining results, both negative and positive, discussions, trial and error, peer reviews, and other time-consuming steps. At Love Canal, more than once when preliminary health studies finally were performed, the people at risk, the mass media, and then the general public were alerted, and often aroused. Those in the limelight simply could not give the scientists time for careful, reflective work. Policy decisions too often had to be made immediately—often under the glare of television camera lights and in front of angry people.

Given the uncertainties in the situation, and given the trust the public is thought to have in science, scientific studies were nevertheless offered as the reasons for policy decisions, in order to enhance the acceptability of the decisions. However, the decisions were necessarily based on other considerations—because major policy decisions about health and welfare are linked to available resources and because, in fact, the scientific findings simply did not exist at the time policy decisions had to be made.

Another complicating factor was added during the second Love Canal year. The New York Department of Health had not publicized and consistently refused to share the findings from the large-scale epidemiological study conducted by that agency in the summer and fall of 1978. Thus, studies such as Paigen's, done with little funding and intended to be pilot or exploratory works, assumed an importance far beyond anything anticipated by their authors, who well knew and often explained the limitations on the data, findings, and conclusions they presented.

In the end, reviews of the few publicized scientific studies were done, and the reviews became instruments of punishment for those who interfered with the emerging official description of the Love Canal problem. The lone voices of those seen as dissenters, were drowned out by the chorus of official explanations of what happened at Love Canal and how the problem was handled (see Levine and Levine 1977).

Autumn 1979 and the Beginning of Winter

Labor Day 1979 was not a happy holiday for the Love Canal community. During the summer, construction crews had worked long hours covering the canal with impacted clay, hurrying to get the huge job done within time limits set by the federal and state funding agencies (see Clement Associates 8/20/80). The fumes from the canal work filled the neighborhood. Although many children spent their weekdays in day camps and day-care centers, some adults began to sicken. As the workers began to put in more overtime hours to finish the job, people feared for their families' health even more than usual.

In late August, when the fumes were hanging in the oppressively humid air, the acting on-site state task force coordinator (who had replaced Mike Cuddy) heeded the organized residents' pleas. With his superior's approval, he authorized a few families to move to a local motel for forty-eight hours. The move was in keeping with the state's contingency plan, approved by a state court judge in June 1979, when the Homeowners Association had sought an injunction against the construction. Although the judge denied the injunction, he accepted the state's plan for moving people out temporarily if they grew ill from fumes or other construction-related causes.

Once the first few families moved out to motels and reported themselves free of headaches, able to sleep nights, not wheezing and coughing, a few other families followed suit, and then more and more, until there were more than seventy families in motels—for forty-eight hours only. The first forty-eight hours ended on August 31. At that point, the state task force director reminded the people that they could not stay in the motels any longer without doctors' certificates. It was not easy on that day, or in the weeks ahead, to get certificates. August 31 was the Saturday preceding Labor Day, and even nonvacationing physicians were difficult to find. Some people had no family doctors. Some Niagara Falls physicians did not believe that Love Canal was related to the state of their patients' health; some feared involvement with pending lawsuits; some had been annoyed when dealing with the DOH during the preceding year and wanted to hear no more about the whole situation.

The other details of that Labor Day weekend are too numerous to present here, but the certification problem indicates one small part of what it meant for the families to be put into temporary quarters. Within a few weeks, six young interns from nearby Erie County lent their time and expertise to the effort, and Niagara Falls physicians certified many of the people. Before the entire episode of motel living was over, however, the Love Canal families had to move from the motels to a private-school dormitory in order to make way for tourists who had Labor Day reservations in the motels; then the families were housed for a few days in a church, to make way for the children returning to the private-school dormitory; then they returned to motel rooms. Finally, after almost two months adrift, more than 125 families returned to their homes.

The drawn-out episode of motel living drew a good deal of public attention, with articles appearing daily in all local papers about the Love Canal "refugees." Jane Fonda, an actress known for her support of environmental causes, made a well-publicized visit to the Love Canal neighborhood and, in tears, expressed her view that the people should be evacuated (Vogel 10/4/79; *Niagara Gazette* 10/4/79). A local editor wrote that Governor Carey's comments that New York State had no funds to do anything further for the Love Canal residents was "Not an Answer" (*Niagara Gazette* 10/11/79). In the second week of October, front-page Niagara Falls and Buffalo news articles (Lynch 10/14/79; Shribman 10/14/79) described the report of a House subcommittee, which said: "A crisis of confidence has arisen with regard to the response of the State of New York to the situation at the Love Canal" (USHR 9/79, p. 20).

Thus, the episode was not only disruptive to the families and financially costly to the state, at over $100 per day per family, but it also further tarnished the image of a caring government. On October 17, Governor Carey suddenly announced his support for a legislative bill introduced by state Senator John Daly and Assemblyman Matthew Murphy (Silver 9/2/79), as part of their many attempts to help the Love Canal residents (Herman 10/17/79). The residents had barely finished cheering when they learned that actual use of the money was far off. Five million dollars had been found for use at Love Canal from unspent state bond monies for urban-renewal projects authorized under laws passed in 1958 (N.Y. Laws 1979, Ch. 732). However, spending those funds depended on finding more funds from federal sources. Five million dollars was less than a quarter of the amount necessary to purchase the homes of those who wanted to move out; and the money also had to be used to "stabilize and revitalize" the neighborhood. Moving people out was intended to be just one part of that effort. Moreover, for the next six months, the local and state officials assigned to the task—headed by Mayor O'Laughlin of Niagara Falls—did not devise an appropriate entity to administer the use of the money. Very soon, many Love Canal people thought of the governor's promise and the new law as little more than symbolic pacifiers intended to keep them quiet until the construction work was complete.

The prospects of getting out of the situation seemed to dim during the last weeks of 1979 and into 1980. The construction work was completed early in 1980, and, as the construction workers left, the state task force thinned down to a skeleton crew for administrative maintenance purposes. To many residents, the huge fence, the locked gates, and the snow-covered, abandoned homes in rings one and two all symbolized their condition: "We're trapped in this cemetery and we'll just be forgotten here." Even the Homeowners Association leaders, usually ready to "put on a cheerful face" despite all odds, seemed wrapped in gloom.

As 1979 drew to a close, however, some events were occurring, far from

Love Canal itself, that would finally clarify and change living conditions for the people. Paradoxically, the understanding of the health problems at Love Canal became more confused than ever, partly as a result of those events.

The Chromosome-Damage Study Begins

On December 20, 1979, the U.S. Justice Department, on behalf of the EPA, filed four lawsuits against the Hooker Chemical Corporation, and others, in the U.S. District Court for the Western District of New York.[1] One action was in connection with Love Canal. In the four suits, the EPA asked for damages and penalties totalling over $124 million (*USA* v. *Hooker* 12/20/79; Molotsky 12/21/79).

In January 1980, when the EPA's Enforcement Division lawyers began to collect evidence for the lawsuits, they consulted with scientists in the agency's Health Research Division. The latter decided that cytogenetic studies (that is, studies of blood chromosomes) of Love Canal residents would be appropriate.

This was not the first time cytogenetic studies had been considered for Love Canal residents. Beverly Paigen had collected specimens for such a study from ten Love Canal volunteers in February 1979. A colleague who had enthusiastically expressed his willingness to lend his laboratory equipment and personnel suddenly withdrew interest after Paigen had drawn the blood samples and prepared the slides. Paigen kept the slides because she continued to think that the studies should be done. Even earlier, Dr. Ernest Hook of the New York Department of Health had proposed such a study but had been refused funding by the EPA on the grounds that the proposal was not specific enough in detail (Williamson 1979).

In January 1980, the EPA scientists discussed the research with a Texas cytogeneticist, Dr. Dante Picciano, who was employed by the Biogenics Corporation, a commercial cytogenetic laboratory. Several EPA scientists had worked with Picciano; in fact, he was one of the experts who had reviewed the proposal submitted to the EPA by Dr. Hook of the DOH. Once Picciano was involved, the EPA asked Dr. Paigen to act as on-site coordinator of the project. She told them she would be most willing to cooperate.

The study of chromosome damage is an expensive enterprise. The long, complex preparation and careful reading of some 200 slides for each individual cost about $250 per person at that time. The scientists decided to start with a small study, the results of which would help them make a decision about pursuing a study with a larger number of subjects. The use of such a pilot project is a standard research technique.

Because chromosome aberrations are estimated to occur usually in only one percent of the population, and because the study groups were to be small, Picciano and Paigen adopted another standard technique for the pilot project. They selected for study a group that could maximize the possibilities of finding valid results. First, they ruled out, for both the study and the comparison groups, volunteers who had experienced major known causes of chromosomal damage within the preceding few months (occupational exposure, Xrays, and viral disorders). Then, to maximize the possibility of finding results, they purposely included in the study people who lived in wet areas, people in homes with air readings showing detectable levels of toxic chemicals, and people who had undergone miscarriages or who had borne children with birth defects since moving to Love Canal. The researchers reasoned that, by comparing what should be a high-risk group to a group from an uncontaminated neighborhood, a decision could be made more easily about whether the amount of difference between the two groups justified a larger study.

Paigen had already selected an experimental group from Love Canal and a control group from a different neighborhood in Niagara Falls when the EPA attorney assigned to the project informed her that the EPA would not pay for the tests for the control group and that the agency would fund no more than a total of 36 cases. Paigen mulled over this unexpected development for more than a week. She had had great difficulty attracting anyone to do clinical studies on the Love Canal people, but she now was faced with the choice of conducting no study at all or a study with no control group. She finally agreed to go ahead, although she knew that the inferences from the results would be limited. The influence of the financial sponsor on scientific research was very clear in this instance.

Dante Picciano, who was responsible for the cytogenetic report, "repeatedly cautioned the Enforcement officials that without contemporary controls, no definite conclusions could be drawn from the analysis of 36 Love Canal residents." Picciano proceeded cautiously. He told his laboratory technician that the slides were from both Love Canal residents and controls. (Picciano 8/15/80).

The Research Report Is Submitted

On May 5th, Picciano wrote to Frode Ulvedal, of the EPA's Health Effects Division that twelve individuals showed chromosome aberrations. (USHR 5/22/80).

The completed report, "Pilot Cytogenetic Study of the Residents of Love Canal, New York," arrived in Washington on Thursday, May 15. Picciano's detailed report to EPA started out with a one-page summary:

> We believe that the results of this pilot study indicate that these residents may have increased frequencies of cells with chromosome breaks and

> marker chromosomes, especially ring chromosomes. . . . In the absence of a contemporary control population, we have estimated that the frequency of individuals with supernumerary acentric fragments to be approximately one (1) in one hundred (100) in normal individuals. This frequency is compared to eight (8) individuals with supernumerary acentric fragments in a total of thirty-six (36) Love Canal residents studied.
>
> It appears that the chemical exposures at Love Canal may be responsible for much of the apparent increase in the observed cytogenetic aberrations and the residents are at an increased risk of neoplastic disease, of having spontaneous abortions and of having children with birth defects. However, in absence of a contemporary control population, prudence must be exerted in the interpretation of such results (Biogenics, 5/14/80).

Picciano's report to the EPA did not seem unusual in format or in its cautious writing style ("we believe", "indicate", "may have", "it appears", "prudence"). He presented the problem, methods, and findings, with several tables and figures showing data, a discussion of relevant literature, his conclusions, and his recommendations. He repeated that the lack of control groups and the small number of subjects in the study limited the inferences to be drawn from the data. However, given the serious nature of the chemical stimuli and the high incidence of damaged chromosomes he found, he expressed the opinion that a larger study was warranted. He said that such a study should include as many Love Canal people as possible—at least fifty from areas of low, medium, and high levels of exposure to chemicals and fifty in a control group, for a total of at least two hundred subjects (Picciano 5/14/80; see also Picciano 1980).

First Responses to the Report[a]

Within 24 hours after the Biogenics Corporation's written report arrived in the EPA offices on May 15, 1980, the document was embroiled in a mix of science and politics in which politics predominated. The general pattern—ignoring the problem, denying responsibility for what can no longer be ignored, rushing in impulsively if it seems expedient to do so, and then minimizing—was set into motion again at Love Canal, this time by the federal government.

This time advisers to the president became more visibly involved than in the past. The White House had known about the Love Canal situation at least since August 1978, when Governor Carey had obtained a declaration

[a]The materials for this section and the next were derived from interviews with Lois Gibbs, Lewis Golinker, Jane Hansen, Vilma Hunt, Congressman John T. LaFalce, Beverly Paigen, and Dante Picciano; from news accounts by Richard Severo (5/27/80), David Shribman (5/25/80), and David Lynch (5/25/80); from a report of the House Subcommittee on Oversight and Investigations (USHR 5/22/80); and from my field notes and observations—in addition to the sources cited specifically throughout.

of emergency for the area from President Carter. On February 21, 1979, Robert Morgado, Governor Carey's chief of staff, sent Jack Watson, President Carter's assistant for intergovernmental affairs, a five-page plea for the federal funding that the New York officials thought had been promised in the summer of 1978 (Morgado 2/21/79). Nine months later, Congressman LaFalce sent President Carter a six-page letter detailing the disorganized, inadequate response of the federal government to Love Canal and asking the president to develop a "responsible, comprehensive plan" for the federal government, both at Love Canal and in environmental disasters in general (LaFalce 11/9/79). Lois Gibbs had written repeatedly to President and Mrs. Carter and had spoken with and written to Jack Watson.

In April 1980, alerted by a request from Mayor O'Laughlin of Niagara Falls for the use of apartments no longer needed for air force personnel, Jane Hansen, an aide to Jack Watson, began to collect information about the health studies done on Love Canal residents. By May 12, she not only had learned about the chromosome damage study but also had discovered that someone was going to tell the newspapers about it. A series of meetings took place that week. In the early afternoon of Friday, May 16, a meeting was held in the White House, chaired by Jack Watson. Other White House staff members were present, as well as representatives from the EPA, the Department of Health and Human Services (HHS), the Federal Emergency Management Agency (FEMA), the National Institute of Environmental Health Sciences (NIEHS), and the Department of Justice.

The participants had many variables to ponder in the course of that meeting. Two participants recalled that the major focus of concern was the health and well-being of the Love Canal residents. Later, the newspapers reported that the White House staff people were interested in using the latest Love Canal developments to win votes for President Carter in Western New York (see MacClennan 6/18/80). According to one participant, however, although electoral politics were discussed in the course of the meeting, they were simply dealt with as one of the factors that could be considered in making a policy decision.

Some of the president's advisors expressed the opinion that, if the Love Canal people were in a situation of health peril, they should be moved from the neighborhood. There was doubt expressed that this expensive and unusual course could be justified by the scientific studies at hand, however. The problem was raised of setting precedents for governmental behavior in future disasters. The decisions finally made at that meeting were that the chromosome study would be reviewed immediately by a panel put together by Dr. David Rall of NIEHS. The question of the appropriate actions to take immediately was postponed, pending the results of that review, which was to be completed no later than the following Wednesday, May 23. Because no massive health-study results were available, the condition of the Love

Canal residents' health would have to be assessed by using whatever small studies were on hand and by doing further health studies. After more extensive studies were done, further decisions could be made about longer-lasting solutions. Little consideration was given at that time to the residents' feelings about participating in further health studies with federal agencies, after their frustrating experiences with the state health studies.

After the White House meeting, Dr. Dante Picciano received a call from EPA officials, informing him of the Rall review panel. According to Dr. Picciano:

> I told EPA officials that the review was fine with me. I then discussed the HHS committee with officers of the Biogenics Corporation. I expressed some concern that there might be a conflict on the part of one committee member because it had been reported that he was investigating the possibility of starting his own cytogenetic analysis company similar to ours. We informed EPA officials of our concern and told EPA that, if that member was removed from the committee, Rall could add anyone he wanted and, in addition, select any one of five individuals we recommended. (Picciano 8/15/80:756)

The review of Dr. Picciano's work will be discussed later.

The Report Is Leaked and Becomes News

There had been general agreement among the participants in the meeting with Watson that the potentially explosive revelations about chromosome damage should be kept confidential until the review of the Picciano study was completed. However, shortly after that meeting, the rumor that insiders had expected swirled through the offices and agencies in Washington. It was confirmed and relayed to people in other parts of the country, too. Someone had "leaked" the Picciano report, with its statements about possible chromosomal damage, to *The New York Times* and the *Buffalo Courier Express,* and the papers planned front-page stories for the following morning.

That afternoon, Beverly Paigen received a call at her office in Buffalo. At first she thought it was just one more in a series of telephone conversations she had already engaged in over the preceding two days with the EPA scientists. They had been discussing the cytogenetic report and the possibilities of conducting further studies, as Picciano had recommended, to try to verify or disprove his conclusions. There had been little that was unusual in the conversations, as far as she could discern, except for the exceptional interest displayed by the White House staff. In mid-afternoon of May 16, however, one of the Washington-based scientists alerted Paigen to the newest development—that the whole story was about to appear in the newspapers on the following morning. Neither Paigen nor the person she spoke with thought that the results of a limited pilot study should be released at this time. The limited, hedged conclusions could increase public uncertainty about scientific reports, and the study certainly would not help provide good scientific evidence for a lawsuit.

Paigen insisted that the people who had participated in the study absolutely must have the results explained to them prior to any public announcements. Her advice prevailed, and a press conference was scheduled for the following morning rather than taking place immediately, as someone at EPA was considering.

In Washington, a White House aide telephoned Congressman John LaFalce with messages that surprised and even startled him. LaFalce was notified that there would be a press conference the next day, that an announcement might be made at the conference that more than 700 families would be moved temporarily from Love Canal, and that LaFalce was to tell neither his staff, the New York senators, the Governor, nor the health commissioner about the press conference! In fact, he was told that plans were afoot to have the press conference take place under the direction of Association Homeowners president Lois Gibbs. Soon, however, LaFalce insisted that the press conference be held in his office in Niagara Falls, with appropriate state and federal authorities notified.

Later in the afternoon of May 16, Lois Gibbs received a telephone call. An EPA official asked her to notify the thirty-six participants in the chromosome study that, within eighteen hours, on the following morning—a Saturday—they would be told the results of the cytogenetic study. Gibbs's mother stayed with the young Gibbs children while Lois Gibbs, Marie Pozniak, and Grace McCoulf spent the next eight hours telephoning and tracking people down, telling them to come to the association headquarters in the morning. They made their final telephone call at 2:30 a.m., to the consternation of the sleepy recipients.

Getting the News

A few hours after the association workers made their last phone call, the early plane arrived from Washington with nine EPA officials aboard, including physicians, scientists, and members of the legal staff. They planned to meet for half-hour intervals, in teams of three, with each person who had participated in the cytogenetic study in order to explain the results.

Despite the haste, they were already too late. By 7:00 a.m., well before the plane left Washington, the news of the study was featured on the television and radio in Niagara Falls. *The New York Times* carried a front-page story, dated the previous day, with full details about the number of affected individuals and the meaning of chromosomal damage (Molotsky 5/17/80). Both local morning papers carried the full story on page one (Moe 5/17/80; *Niagara Gazette* 5/17/80). Once again, a dire health warning was announced by the news media before the affected people learned about it on an individual basis from informed officials.

When the EPA officials arrived, the thirty-six families were waiting for them. Inside the Homeowners Association building, small groups of families clustered anxiously, awaiting their turns to hear whether they were in the group in which chromosome damage had been discovered. By 11:00 a.m., several families had heard bad news. One man in his late fifties wiped away his tears as he described how he carried out his parental responsibility, when he learned that he was one of those with aberrant findings. Like the others, he had been told about the possible implications for genetic damage, fetal damage, and the development of cancer related to chromosomal damage. He telephoned his twenty-seven-year-old married daughter.

> My daughter works and she's been on the pill for five years. A few weeks ago, she and her hubby decided they've got enough of a nest egg now, and they can think about starting a family. Well, she was born here, and I thought I better let her know about my blood and all. So I called her up long distance.
>
> When we started talking she was so happy to hear from me, her voice was all light and happy, and by the time we finished talking, her voice was so low—I think she was almost crying. . . . Gee, I sure hated to have to tell my little girl that kind of news.

At noon, while the subjects of the study were still being ushered into the rooms with the EPA teams, press conferences were taking place simultaneously in the EPA headquarters in Washington and in the Niagara Falls office of Congressman LaFalce. Not until a quarter-hour before the conferences began were Governor Carey and Commissioner Axelrod informed of the forthcoming announcements—a slight neither would soon forget. A lawyer for Hooker had been notified in time to attend the press conference, however (Molotsky 5/18/80).

In Washington, Barbara Blum, EPA's Deputy Administrator, issued a press release stating that eleven of thirty-six persons tested at Love Canal had shown chromosomal damage. The study, it was reported, was "*now* under intense review by recognized expert geneticists" (emphasis added), and, once that review was complete, decisions would be made as to whether the new evidence, added to other information about health and environmental effects, would justify a recommendation for temporary relocation of Love Canal residents. Blum alluded also to "a larger, more comprehensive sampling and testing program and contingency plans in the event that temporary relocations of 'impacted area' residents is shown to be needed" (USEPA 5/17/80; see also Moe 5/17/80).[2]

The Hooker Company representative present at the Niagara Falls press conference urged prudence in interpreting the study results. There were far more intense reactions from others, however, who learned that policy decisions would be deferred until at least the following Wednesday, five days away. Mayor O'Laughlin, summoned hastily for the press conference, angrily told the regional EPA director: "You've thrown a bomb and now you're going to fly out leaving us holding the bag" (MacClennan 5/18/80). The residents were furious at the delay. *"Get us out now!"* shouted Pat Pino at the federal officials (MacClennan 5/18/80). Pino's reaction reflected the sentiments of many Love Canal people.

The press reaction was immediate and overwhelming. Within hours, the news was on front pages all over the world. Within two days, the Homeowners Association building and lawns, and the surrounding streets were so packed with reporters and cameramen that one local television station devoted a story to that aspect alone.[3] So many telephone calls came in, from all over the United States and foreign countries, that the association had to install additional telephone lines immediately.

The warning in Blum's press release, that "the chromosomal study . . . cannot be regarded as a complete and definitive scientific study" (USEPA 5/17/80), disappeared in the storm of attention and the newly aroused fears and hopes at Love Canal.

No matter what the circumstances were that put the EPA into the limelight at this time, the parallels between the behavior of that agency and that of the New York DOH two years earlier were striking. High-level government officials from an agency devoted to the protection of people's health and safety made authoritative-sounding pronouncements about serious health hazards for Love Canal people. They claimed that their conclusions were based on preliminary scientific studies. They had no plan, derived from formulated policies ready at the time of the public announcements of the danger, to announce also some beneficial, sensible-sounding measures to assist the affected people. No one with real authority and correct information was on hand and available in Niagara Falls after the announcement was made, to explain the highly technical, ominous-sounding matters in further detail. Agency people merely stated what, as sympathetic and well-intentioned individuals, they thought *should* be done and *possibly* would be done *if* there were resources and *if* forthcoming scientific statements and further studies showed that definite links could be made between residence at Love Canal and manifestations of physical disorder—a task that the Love Canal residents, as well as the most learned scientists, knew to be impossible, given the condition of scientific knowledge about toxic chemicals and their chronic and long-term effects.

The EPA officials had not anticipated the intensity of people's reactions, the impact their pronouncements would have on the mass media, nor the unwavering scrutiny that their every statement and action would now undergo, even though they might have turned to the New York DOH experience for guidance. Working scientists and lower-level staff and public-relations people arrived on the site to set up and staff offices (for EPA and FEMA, and later CDC), ignorant of the concerns of the residents and without authority to make decisions on their own. Like their counterparts from New York State, the federal personnel learned a great deal very quickly, through sometimes humiliating experiences acted out in very public places.

When we compare the two groups, however, we see that the New York State people had had the advantage. They had arrived while the residents still had feelings of trust and goodwill toward governmental helping agencies. The federal people reaped the harvest of anger sown not only by their own but by their predecessors' behavior. Ed Pozniak summed it up: "I thought the feds and the state guys weren't even talking to each other. It looked to me like the feds took lessons from the state! It's like August '78 all over again!"

It was not surprising that the familiar question, "Who pays the bill?" surfaced immediately, for it had never been resolved. The issue of fiscal responsibility is, to say the least, confused and complicated in our tripart, multilevel, electorally sensitive, pressure-prone governmental system. This complicated structure forces its participants into complex negotiations between and among local, state, and federal branches and agencies. Complex negotiations require compromises, trade-offs, and time—none of these necessarily leading to swift, resolute, and satisfactory solutions to new, pressing problems.

Despite the many similarities between events in the spring of 1980, and the situation in August 1978, there were some important differences. Twenty-two months of experiences and events had intervened. The organized Love Canal residents were no longer naive, no longer trustful, and no longer unsure of how to proceed.

During the first year of their ordeal, they had been alerted by state agencies to the hazards they faced and then, in their view, had been given insufficient help. Since the autumn of 1979, they had watched the ineffectual attempts of municipal and county representatives to make real the promise—inherent in the $5 million in revitalization funds—to help solve the problem of their getting *out*. They felt that sufficient federal help had not been appropriately forthcoming from the very beginning, and no reasons had been offered that satisfied them as taxpayers and citizens in need.

The twenty-two months of accumulated experiences resulted now in a speeded-up quality in some ensuing events. Matters that had taken weeks or months to develop between the state government agencies and the residents in the spring, summer, and fall of 1978 took hours or days at this time, as the residents encountered the federal authorities' attempts to join the action at Love Canal. The residents had grown adept at assessing situations, at exerting behind-the-scenes pressures, and at dramatically presenting their stories and goals to the world. They well understood that it was of prime importance to present their definition of the Love Canal problem and to try to make it the controlling one in this situation. By now, there was little question that many more of the Love Canal people were very angry. The EPA, the federal government, and the world were to get that message very shortly.

"We Are Holding Two EPA Officials Hostage"

On the morning of Monday, May 19, two days after the first group of thirty-six residents had been told the results of the chromosome test, a small crowd surrounded the association headquarters again. Members of the press were milling about, scenting the newest developments. This time, some residents were hearing or waiting to hear the results of another examination, this one for peripheral nerve damage, which had been conducted by Dr. Stephen Barron, a neurologist from SUNY/Buffalo and the Veterans Administration hospital at Buffalo.

The nerve-conduction study had not been undertaken in connection with the EPA's chromosome study, but the subjects for the study had been sought at the same time. Barron's study was a pilot project done in order to apply for a research grant if further study seemed warranted. He had found that a higher percentage of the thirty-five Love Canal subjects showed slowing in conduction of nerve impulses when compared with twenty people in a control group who did not live near Love Canal.

Dr. Barron had included a summary statement about the results in an abstract that was part of a grant application to the EPA. On that busy Friday, May 16, he was surprised when he was telephoned by someone from EPA, asking whether the agency could have access to his data. The caller explained that the agency wanted to review all health data known about the Love Canal people (USHR 5/22/80, pp. 47-48). Barron wanted to be helpful and to cooperate with an agency that might fund his proposed study. He agreed to meet with the EPA scientists on Saturday, May 17, to submit his research design to their scrutiny. He refused to share actual numbers with them, however, for at that time he planned to take several weeks for statistical analysis and then draft careful personal letters to each of the participants in the study, telling them about the test results. He flew to Washington for the urgent Saturday meeting, explained his research design, and flew back to Buffalo. He spent Sunday working on the data, because he now believed that he must talk with the residents who had participated in the study before his pilot project, too, was leaked. By Monday morning, he was ready in a room at the Homeowners Association headquarters, where he could meet privately with the families. He met with them throughout what became a day filled with unusual events. He presented his findings in a cautious manner, explaining that further study would be essential before anything could be *conclusively* determined about the meaning of his results. We will return to Dr. Barron's study later.

In addition to the families waiting for test results, a few dozen other people were gathered at the association building. Everyone wanted to know what was going to happen next. Would there be more chromosome studies? Would there be health studies of other kinds? Should they move right out? What should they tell their married children? Should they have another child? Could someone please repeat the meaning of the chromosome study—and perhaps repeat it more than once, so that they could really understand what it all meant? They had other questions as well.

Two EPA representatives from the Washington office had remained in the city to be available as necessary, but they were not present at the Love Canal headquarters where residents were gathering. In the excitement of the weekend, no one had written down the name of their hotel or their telephone numbers. They were evidently busy elsewhere, for they did not make themselves available to the residents during the long morning, when frustration and anxiety mounted among the waiting people. The scene reminded Gibbs of the panic she had faced upon returning from Albany after the state health commissioner's announcement of August 2, 1978. Annoyed now at the government officials, and wary of the direction the gathering crowd's anger might take, she systematically telephoned the numerous hotels and motels in the Niagara Falls area, attempting to locate the two EPA officials and to urge them to come to the association headquarters immediately. She had no luck for several hours, however.

By early afternoon, the crowd of about fifty, now mostly women and young children, was growing restless, angry, and frustrated. At about 1:00 p.m. a newspaper headline stirred the crowd: "White House Blocked Canal Pullout (MacClennan 5/20/80*a*). Upon reading the headline, a woman walked out into the street and stopped a passing car. She told the driver he could not drive through the streets. Another woman joined the first one in the street and together they stopped traffic. One woman diverted the watching crowd when she dashed across the street to a boarded-up house and poured gasoline on the front lawn in the form of the letters, E,P,A, and set the grass on fire. The blaze drew a hearty cheer from the crowd. The crowd swelled when school buses returned in the afternoon and the schoolchildren arrived, including teenagers. Soon several policemen appeared, directed traffic away, and then stood to the side, quietly making their presence known in this increasingly tense situation.

At last, at about 3:30 p.m., the EPA officials arrived; Frank Napal, the public relations officer, arrived first, followed by Dr. James Lucas. When the two men walked into the association office, Gibbs told them they were going to be detained there to protect them from the angry crowd, who wanted them held until word came from Washington that Love Canal people would be moved out. Gibbs then telephoned the White House and firmly told the switchboard operator: "We are holding two EPA officials hostage." She was put on hold! Undaunted by that initial reception, she explained what had happened, emphasizing the widespread anger of the crowd.

The doors to the office and the building were now blocked by angry women and curious children. The EPA officials remained in the office for about five hours. They were fed homemade oatmeal cookies and sandwiches and were allowed to meet with the eager press, who were ushered into the room one by one. There was steady communication between the two EPA officials and members of President Carter's staff and the EPA's Washington office.

By the time the "hostages" were released, at 9:30 p.m., the crowd numbered about 125, and included many men. Observing were some two dozen cameramen and journalists, and standing across the street were uniformed policemen and several men in blue blazers. There were also three men whose well-tailored suits and trenchcoats added a dashing sartorial contrast to the jeans, sweatshirts, and nylon windbreakers of the Love Canal men. ("All them guys must be the Feds!")

Despite the cookies, the women and children blocking doors, and the slight frames of Lois Gibbs and Barbara Quimby who bravely guarded the two EPA members, the event had serious overtones and was taken seriously by the residents and federal officials alike (MacDonald 5/20/80; MacClennan 5/20/80*a*). The two EPA officials were recalled to Washington the next day, and the local U.S. attorney gave a stiff warning to the women most involved in the episode (Barbanel 5/21/80).

Although the EPA spokesperson insisted that the incident would not affect decisions, which would be based on reviews of health data, Gibbs commented: "We've gotten more attention [from the White House] in half a day than we've gotten in two years" (MacDonald 5/20/80).

Before noon on Wednesday, May 21, well before the completion of the Rall committee's review of the report on cytogenetic studies, the EPA announced a decision that had been reached on Tuesday evening, May 20 (*The New York Times* 5/21/80). At the moment that decision was announced, about fifty residents and twenty-five reporters and cameramen were packed on the lawn outside the Homeowners Association headquarters. Lois Gibbs stood on a chair outside the building, a telephone receiver pressed to her ear, repeating sentence by sentence the press release being read at that moment in Washington by Barbara Blum, the EPA's deputy administrator:

> President Carter [upon the request of Governor Carey] . . . declared an emergency to permit the Federal Government and the State of New York to undertake the temporary relocation of approximately 700 families in the Love Canal area of Niagara Falls, New York who have been exposed to toxic wastes deposited there by the Hooker Chemical Company (USEPA 5/21/80).

The press release noted that the government suits against Hooker would "be amended to seek reimbursement for costs expended in this effort" (USEPA 5/21/80).

Immediately after the announcement, the residents' reactions were mixed. Some left hastily to make motel reservations for that night. The leaders of the Homeowners Association, thinking about the hard-fought battle, were jubilant. They had learned to view even minor concessions as victories. Knowing that the financially costly decision just announced would give them another weapon to help reach their goal of permanent relocation, they sensed that the end was near. A less-than-hearty cheer arose from the other residents. They were skeptical. For many, the memories of motel life remained fresh and unpleasant, and the future still seemed uncertain. (See Tyson 6/4/80.)

Reviews of the Chromosome Study

The decision that was announced on May 21 and relayed to the excited throng outside the Love Canal Homeowners Association headquarters and to the rest of the world as well was *not* based on the planned review of the pilot cytogenetic study.

As mentioned earlier, on May 16, Dr. Picciano was informed that Dr. Rall, head of NIEHS, was putting together a review panel to fly to Houston on Saturday, May 17, to review the study. Although the proposed review was unusual in terms of its urgency, the direct White House involvement, and the fact that it was undertaken by a federal agency different from the one he had originally contracted with, Picciano agreed to cooperate fully. He objected, however, to the inclusion of one member of the proposed committee, on the basis of possible conflict of interest. He told the EPA that he would be happy with the substitution of any one of five other individuals on the committee, whom he would recommend. He had thought the EPA was in agreement. However, according to Picciano:

> On Monday, May 19, I received a call from Charles Carter . . . acting for Rall. He told me no one had been removed from the committee, but he would add one of our five choices. . . . [W]e did not object. Later on the same day, I received a call. . . . Not one of our five nominees had been selected. . . . At approximately midnight, Carter called my office. . . .

> I expressed our concerns to him and asked him if he would like to come to the laboratory and discuss the makeup of the committee. . . . He declined to visit the laboratory and discuss the committee. . . . After repeated negotiations through [Barbara] Blum with the HHS committee (now made up of five reviewers and two observers) we were asked to nominate one individual. He was rejected. The HHS committee would no longer negotiate the composition of the committee, and they decided to review our report and make their own report without visiting our laboratory or seeing our results. (Picciano 8/15/80; p. 756).[4]

This episode was described in the newspapers as Dr. Picciano's refusal to allow a federal review panel into his laboratory (MacClennan 5/20/80*b*). He reportedly told a *Buffalo Evening News* reporter: "We have cooperated with EPA, and we are happy to continue to cooperate, but we don't like the idea of this panel. Some aren't geneticists [5 of 8 were not], some aren't from EPA, some are coming from the state Department of Health" (MacClennan 5/20/80*b*). In any case, the committee that flew to Houston did not go to the laboratory and did not examine slides or other primary data; it simply reviewed the report Dr. Picciano had submitted to EPA. The committee decided that it was unable to evaluate the report fully and that there was "an inadequate basis for any scientific or medical inference from these data" (Bender 5/21/80).[5]

Between May 21 and June 12, there were five more reviews of the cytogenetic study. In the ensuing controversy over the quality of the study and the firmness of the conclusions that could be drawn, all parties agreed on the deficiencies created by the lack of contemporary control groups—first stressed by Dr. Picciano. The reviewers differed on matters of the preparation of the materials and the identification of the damaged cells—important technical differences about which there is no total agreement among cytogeneticists. The various reviews will be described in a later section of this chapter.

Alarums and Retreats

While the scientists were reviewing the pilot study of chromosome damage, the Love Canal residents anxiously lived through the now-familiar steps three and four in the "ignore, deny, sound alarms, retreat and minimize" pattern of governmental intervention.

The day after the public announcement that more than 700 families could be temporarily evacuated from their Love Canal homes, Dr. Stephen Gage, EPA's assistant administrator for research and development, responded to questions from a House subcommittee by saying that the EPA had felt doubtful about the blood chromosome study from the very beginning, because of its obvious limitations [USHR, 5/22/80, Gage testimony (p. 25ff.) Shribman, 5/22/80].

His publicized statements confused many Love Canal residents. With the news that they had suffered chromosomal damage and possible genetic harm (Moe 5/17/80; Molotsky 5/17/80), they had felt sure that solid scientific tests had shown that they were in grave danger. With the federal government at last involved in the health issue, they were sure they would be moved out, permanently or temporarily, while more complete studies of what had happened to their health were carried out. (State/Federal Information Bulletin, 5/26/80).

They were soon told, however, that plans to buy their homes permanently were contingent upon tests of *both* health and environment (MacClennan and Shribman 5/22/80). In less than a week, there was a turnabout. On May 27, a meeting was held between officials from EPA, the Federal Emergency Management Agency (FEMA), which replaced the FDAA, the governor's task force, and community leaders. Gage telephoned the group from Washington with the news of the EPA's latest decision: Although the EPA would fund further medical studies, the EPA would not take on any responsibility for permanent relocation of residents, and they would not even recommend any actions to anyone—neither residents nor the president—based on their health-study findings. By that time, in fact, the Office of Management and Budget had made it clear to a congressional subcommittee that the EPA resources for health studies would be so limited that long-term health studies would be out of the question for that agency (USHR 5/22/80, p. 30). (The studies were assigned to the CDC.)

Gibbs told a reporter after the meeting, regarding the relationship between health effects and policy decisions:

> Then it doesn't matter if there's evidence of 100 percent birth defects or 100 percent death. . . . It's all a political ball game and the sooner we all realize that the better. We'll get moved depending on how many telegrams for re-election President Carter gets. (Billington 5/28/80).

The shock waves from the newest EPA announcement reverberated throughout the community. The various organized citizen groups at Love Canal decided to refuse to cooperate with any further health studies until the decision to move them out was clarified (Barbanel 5/28/80). The Love Canal Homeowners Association added the proviso that they would not cooperate unless they and their chosen consultants could participate in every step of the planning and implementation of health studies, from hypotheses formulation through dissemination of the final conclusions.

As it turned out, however, specific studies of health did not serve as a basis for the decisions regarding either temporary or permanent relocation. The activity concerning Love Canal shifted to Congress in late May (Shribman and MacClennan 5/28/80). Senators Javits and Moynihan and

Congressman LaFalce worked feverishly on the problem (see Javits 7/2/80; Powell 6/14/80). The agreement enabling the purchase of the Love Canal residents' homes was finally made in late August 1980, when the FEMA representatives worked out a financial arrangement with representatives from the state interagency task force. In October, one month before election day, President Carter and a host of other political figures made a whirlwind tour of the Buffalo-Niagara Falls area, to capitalize on the president's signing of the appropriation measure that would allow the home purchases and some other necessary cleanup measures in Western New York. Six months after that exciting political drama, the proposed health studies for Love Canal residents were officially canceled by the new federal administration (*Buffalo Evening News* 4/1/81).

It was clear that specific health conditions of Love Canal residents were not important factors in the policy decisions made at that time. The decisions were made *before* the reviews of preliminary studies and before more detailed studies were done. The pressures from citizens, to be described later, the unceasing publicity about Love Canal, and a president and political party trying to stay in power all played influential parts.

Further Reviews of the Chromosome Study

Five reviews of the chromosome study were submitted to the EPA.[6] The first, by the committee impaneled by Dr. Rall, was described earlier.

The second took place at Picciano's request when the chief of the Toxic Effects Branch of the Health Review Division of EPA, Dr. Sidney Green, examined the slides in the Houston laboratory on May 20. Dr. Green called the slides "of good quality" and confirmed that the techniques used "indicated no deviation from established protocols for determining cytogenetic aberrations in lymphocytes . . . the fact that supernumerary acentric fragments and ring chromosomes were observed is biologically significant" (Green 5/28/80).

The third review was done at Picciano's request by a group headed by Dr. Jack Kilian, professor of occupational medicine at the University of Texas Health Science Center.[7] They examined the slides at the laboratory and supported Picciano's methods, findings, and conclusions, including his warnings that the lack of a control group limited the possible inference to be drawn from the study (Kilian 6/5/80).

A fourth review during that period was undertaken by a committee of seven headed by Dr. Roy Albert of New York University's Medical Center.[8] That group met in New York City, read Picciano's report, studied photocopies of photographs of the slides and arrived at the following conclusion:

> Due to technical deficiencies, including lack of controls, lack of examination of the chromosomal preparations on a blind basis and non-optimal culture conditions, the Biogenics Corporation study should be regarded as indeterminate. (Albert 6/12/80, p. 6)

Despite the indeterminacy of the study, they were able to decide:

> The frequency of chromosomal abnormalities reported by Biogenics or observed by the panel was considered to be well within normal limits. (Albert 6/12/80, p. 5)

The fifth study was done, at Picciano's request, on June 12, 1980, by Dr. Margery W. Shaw of the Medical Genetics Center at the University of Texas. Shaw examined photographs of the slides Picciano had studied. She agreed in general with Dr. Picciano's findings and his conclusions and limited inferences (Shaw 1980). In a letter to *Science*, Dr. Shaw, cognizant of the highly politicized situation, stated that she had not met Picciano until the day he asked her to review his work. Referring to the Albert report, Shaw questioned that committee's conclusions, pointing out:

> The results are neither positive nor negative because of absence of contemporary controls. I find it difficult to understand why the EPA panel stated flatly that the absence of simultaneous controls was a very serious deficiency of the study and then stated that Picciano's results were considered to be well within normal limits. (Shaw 1980, p. 751)

In sum, then, of the five reviews of Dr. Picciano's study submitted to the EPA, two were requested by the federal government and three were requested by Dr. Picciano. All reviewers agreed with Picciano's cautious statements that, because contemporary controls were lacking, the conclusions and inferences to be drawn were limited. Support for the study's methods, findings, and conclusions—or lack of support—correlated with either of two factors: the source of the request for the review or the information looked at. Neither of the two government-requested review groups examined data or entered the laboratory. Instead, one reached negative conclusions by relying entirely on the report, and the other also examined photocopies of photographs of slides. Two of the three other groups, who all reported more favorable conclusions, examined the slides, and all visited the laboratory where the work was done, in addition to reading the report. It is striking that only the two reviews that were *not* supportive of Dr. Picciano's conclusions were given widespread publicity. On June 12, the day Dr. Roy Albert submitted his committee's largely negative review to the EPA, a *New York Times* editorial appeared inveighing against Picciano's study, calling it a "gobbledygook Bomb at Love Canal" (*New York Times* 6/12/80).

A scathing attack on Dr. Picciano's study, and on Picciano himself, appeared in the journal *Science* on June 13, 1980. The title of the piece reflects the tone: "Love Canal: False Alarm Caused by Botched Study." A selected quotation under the title stated: "In the opinion of many experts, the chromosome damage study ordered by the EPA has close to zero scientific significance" (Kolata 1980). The article discussed only the two nonsupportive reviews of Picciano's report and made no mention of the two positive reviews available at that time.[9]

In the ensuing storm of publicity, Dr. Dante Picciano became the scapegoat, drawing off official guilt and official anger for official bungling. For doing a blood analysis his company was under contract to do, his reputation was damaged, and the episode put such a strain on his position in the company that he left his job there (Picciano, private communication, 3/29/80). The Love Canal residents spoke of him as an unwitting martyr to their cause.

The publicity had consequences for the Love Canal residents as well. Lois Gibbs called Dr. Clark Heath of the Centers for Disease Control (CDC), who calmly told her and several residents (who had taken part in the study) that Picciano's work could not be supported scientifically. One woman told Dr. Heath that she had planned to have her tubes tied and asked him how she would have felt if she had had it done the previous day. The residents demanded in vain that CDC send them a letter stating their chromosomes were not damaged, a letter comparable to the one each had received from EPA. Heath refused.

One consequence of the way the whole episode was handled by the federal agencies, from the urgent announcement of May 17, 1980, to the bland reassurances of June 13, was that the relationship between the federal agencies and the residents deteriorated. What the residents perceived as a lack of sensitivity to their needs and feelings only deepened the hostility many of them now felt toward the EPA, and their angry feelings were directed toward the CDC and the FEMA as well. What seems surprising was not that the Love Canal citizens called for a boycott of further government-sponsored health studies (Tyson 5/27/80; Barbanel 5/28/80; Love Canal Homeowners 6/16/80),[10] and not that they shouted at a group of well-respected and well-meaning CDC physicians who tried to explain the meanings of chromosomal damage to them in a public meeting on June 21, 1980, and not that they were abrupt with EPA and other agencies' personnel at times, but that the people of Love Canal continued to communicate with the federal agency representatives, as well as with state personnel, throughout the remainder of the summer of 1980.

There was a second consequence of the on-again/off-again chromosome interpretations. Now the Love Canal residents were more convinced than ever that they were not being told the truth. They had accused the state of a cover up, and now the federal agencies seemed involved, too (*Buffalo Evening News*, 6/21/80). Many other people were wondering what the truth really was about the Love Canal residents' health. In the midst of the confusion, Governor Carey appointed a panel to cut through all the ambiguities about health effects, ambiguities created in part by scientific controversy and in part by the political context of policy decision making.

Rolling Out the Big Wheels

Public attention was now drawn to legal proceedings pertinent to Love Canal. On April 28, 1980, New York's attorney general filed a lawsuit against the Hooker Corporation, seeking $365 million (MacDonald 4/28/80). On May 19, Hooker filed its denials and affirmative defenses to the federal lawsuits that had been started five months before—in December 1979. In the midst of the latest uproar at Love Canal, legislative hearings on toxic waste disposal continued, with Love Canal residents testifying before a Senate subcommittee in June (U.S. Senate 6/6/80; Lynch 6/8/80).

Just what the health effects were at Love Canal, and how valid and reliable the studies were that had been done by the New York Department of Health, by Beverly Paigen, by Dante Picciano, by Stephen Barron—or by anyone else—were matters of importance not only to residents, to scientists, and to the public in general but specifically to the governor, to the health commissioner, to New York's attorney general, to the U.S. Justice Department, to the Congress, to the Hooker Chemical Corporation, and to the chemical manufacturing and disposal industries.

On April 27, 1980, State Senator Thomas Bartosiewicz wrote to Governor Carey that after conducting an investigation he was "persuaded that residents of the Love Canal area are the victims of a cover-up involving public and private entities including the State's Department of Health" (Bartosiewicz 4/27/80). On May 17, the chromosome breakage story put the Love Canal back on the front page of newspapers all over the world. On May 25, 1980, interviews with Paigen and Commissioner Axelrod, airing their differences, comprised a segment of CBS's "60 Minutes." The *New York Times* carried a front-page story raising questions about the work of the Department of Health (Meislin 5/25/80, 6/9/80). On May 28, Senator Bartosiewicz wrote to all New York State Senators asking them to cosponsor a resolution to the governor that a Moreland Act investigation (an independent commission with subpoena powers) be commenced. It was in this complex setting, which included attacks on the Department of Health that Governor Carey acted.

Rather than establishing an investigatory commission under the Moreland Act,[11] as state Senator Thomas Bartosiewicz requested, the governor issued Executive Order #102 on June 4, 1980. The governor's order established a temporary panel—"Panel to Review Scientific Studies and the Development of Policy on Problems Resulting from Hazardous Wastes." Among other things, the panel was charged to review and evaluate the Love Canal medical and scientific data and studies prepared by or at the request of state and federal agencies, as well as studies done by any other agency, group or organization. The panel was also asked to review and comment upon the proper relationship between state and federal agencies in local health emergencies such as that at Love Canal. As the governor personally explained to the prestigious panel members, the order was issued because: "In light of recent conflicting and confusing reports of scientific findings at Love Canal, public policy decisions affecting the residents of that area had been made more difficult" (Carey 6/4/80).

In short, the panel was to clear the muddied waters with a clean stream of neutral, objective, evaluative science. According to a health department spokesman, Health Commissioner Axelrod welcomed the review "by an outside and impartial panel" (Shribman and MacClennan 6/5/80).

Dr. Lewis Thomas, chancellor of Memorial Sloan-Kettering Hospital, agreed to serve as the panel chairman. Dr. Saul Farber, dean of the School of Medicine at New York University Medical Center, was listed as secretary for the panel. The other members were Dr. Arthur Upton, former head of the National Cancer Institute and now chairman of the Institute of Environmental Medicine at New York University Medical Center; Dr. Attallah Kappas, physician-in-chief of the Rockefeller University Hospital; and Dr. Richard A. Doherty, an associate professor of pediatrics, genetics, obstetrics, and radiation in the Department of Biology and Biophysics at the University of Rochester Medical Center.

Charged with their responsibility, the reviewers, along with administrative and technical assistants from the New York Department of Health and the Health Planning Commission, all set to work. The Thomas panel met in June and July and issued a final report to the governor and the legislature on October 8, 1980.

The Thomas Panel Report

Unlike so many government-sponsored reports that lie in archives, neglected from birth, to be retrieved only occasionally by a specialist, the Thomas report was a smash hit from the moment it was issued. The report was circulated far and wide and was cited thereafter as *the* authoritative summary of health effects at Love Canal. It was broadcast and published in mass news outlets (for example, Allan 10/11/80; Meislin 10/11/80; Moe 10/11/80; *Time* 11/3/80), featured in editorials (for example, *The New York*

Times 10/17/80; *Wall Street Journal* 10/20/80), and "briefed" in *Science* (Smith 1980) for the edification of *Science*'s audience of more than 150,000 professionals (see also Paigen 1981; Levine 1981; Smith 1981). The editorial in the *New York Times* became the centerpiece for an advertisement presenting the Chemical Manufacturers Association's viewpoint on the proposed "Superfund" regulations of toxic-waste disposal (America's Chemical Industry 11/3/80), and the U.S. Chamber of Commerce devoted a newsletter (circulated to 650 newspapers and magazines and 30 radio stations) to the Thomas panel report (Lesher 11/3/80).

A Hooker Chemical Company spokesman referred to the report to shore up the company's position in an interview the day after the company filed cross-claims and counterclaims in the U.S. District Court for the Western District of New York against New York State, the city of Niagara Falls, its board of education, and the Niagara County Health Department (Hooker 10/28/80; Tyson 10/29/80). For employees and the interested public, Hooker provides an attractive brochure, which consists of little more than quotations from the Thomas panel report (Hooker 11/80), and the Hooker Company president, Donald L. Baeder, referred to the report in an article in a professional journal—to explain "what really happened" at Love Canal (Baeder 1980).

There have been few attempts to examine the Thomas report critically. Short columns were written by two journalists who have a deeper knowledge than most of the Love Canal story (Ember 11/10/80; MacClennan, 10/26/80), but the document has not been inspected closely by scientists. Rather, the summary statements and conclusions have been publicized, with an unquestioned assumption that the report was carefully crafted because it was the work of people of excellent repute.

Because the Thomas panel report has had such wide circulation, has had a deleterious effect on the professional lives of some whose work was discussed in its pages, and may well influence a range of public policies about chemical waste products, it is important to scrutinize here some important parts of the report.

Health Effects

The major questions about health effects stemming from Love Canal chemicals appear to be fully answered in Dr. Thomas's cover letter to Governor Carey, in which he states:

> As a result of this review, the Panel has concluded that there has been no demonstration of acute health effects linked to exposure to hazardous wastes at the Love Canal site. The Panel has also concluded that chronic effects of hazardous waste exposure at Love Canal have neither been established or [*sic*] ruled out yet, in a scientifically rigorous manner. (Thomas 10/8/80)

"Scientifically rigorous" conclusions are generally based on empirical (factual) evidence. Although one of the Thomas report's appendixes contains an annotated list of studies, the report is very different from the conventional scientific review article, which is replete with references to precise locations, in specific documents, where actual data and statements may be found and independently examined by the reader of the article. In the Thomas report there are no precise page references and often no precise documentary reference. Whereas the authors of conventional scientific articles expect and welcome inquiries, the Thomas panel members did not respond to scholarly questions about the factual bases for the report's statements. Instead, Dr. Thomas, speaking for the panel, replied as follows to my inquiries about their sources of information:

> I have consulted with other members of the Panel regarding the numerous requests for more detailed information that we have received from time to time since the release of the report. We have decided that the report should stand on its own without further comment or amplification at this time. (Thomas 11/4/80)

Acute Health Effects

Does the report stand on its own? A close examination reveals first of all that it contains unsupported conclusions about acute illnesses. The panel states:

> It is clear enough from the available data that no *acute* cases of intoxication by chemical pollutants have been observed within any part of the Love Canal community, "wet" or "dry." That is, no clusters of cases of acute liver disease or kidney disease or pulmonary manifestations, or hemolytic anemia or agranulocytosis, and certainly no peripheral or central nervous system syndromes. (Thomas 10/80)

There are twenty-three documents listed in the report's annotated bibliography, presumably the entire empirical basis for the narrative and conclusions in the body of the report. Eleven of the twenty-three studies are concerned with human health. Of the eleven studies, four were described as "in progress" or with "no final report," leaving seven. One study examined 112 construction workers, not Love Canal residents. Three studies were of *chronic* effects (Thomas 10/80). Two others are reports on miscarriages, birth defects, and low birth weights (NYDOH 9/78; Vianna et al. 4/80). Only one citation (Vianna and Fitzpatrick) might possibly have provided some evidence for conclusions about some of the specifically named health effects. The bibliographic notation states: "Twelve clinics were held and about 4,000 blood specimens collected. Blood counts were not unusual. No clinical evidence of liver disease. Some persons were reported as having abnormal liver tests which diminished after relocation (Thomas 10/80).

I wrote to the authors requesting information about their study and a copy of the final report, "Blood Counts and Liver Functions" listed in the Thomas bibliography (Levine 2/19/81, 3/5/81). In reply, Dr. Vianna stated that as of March 12, 1981 (five months after the Thomas report):

> The study referred to by the Thomas Panel is an *ongoing one that will not be completed for several months* . . . we examined the results of blood testing on each individual resident who participated in this investigation. Review of this data suggested that none of the entities . . . [acute liver disease, hemolytic anemia, agranulocytosis, hemotologic abnormalities] . . . occurred with excessive frequency . . . I believe the Thomas Panel adequately commented on the preliminary results of our blood tests (Vianna 3/12/81, emphasis added).

I wrote once again, asking for the specific blood tests performed and asking how the information had been conveyed to the panel (Levine 3/18/81). Dr. Vianna responded that "[t]he specific tests were profile 1 and 2 screening" and that he had conveyed information to the Thomas Panel in a summary memo "dealing with our initial findings" (Vianna 3/24/81). The summary memo contained no data, no statistical summaries, no conclusions, no comments about the outcomes of blood studies, and no information about acute liver disease, hemolytic anemia or agranulocytosis (Vianna 6/19/80).[12]

The panel was also given a brief 1978 DOH preliminary report on some liver-function tests (NYDOH 1979, Sec. 57; Dowling 6/16/80). Although the Thomas Panel did not refer to this report, which was prepared in the fall of 1978, it provides two tables showing liver-function tests before and after evacuation from Love Canal. The sketchy data show that sixteen of eighteen residents whose liver-function tests were in the abnormal range tested normal after moving from the area. The report concluded:

> Analysis of various combinations of liver test abnormalities, corrected for age and sex, suggested that, on the average, the proportion of abnormal liver tests diminished among persons relocated. None had clinical evidence of liver disease.
>
> No unusual patterns of hematologic abnormalities have been detected (NYDOH 1979).

The report has no other information on blood specimens beyond values for liver-function tests for twenty-seven individuals [of the "4,386 blood samples taken from 3,919 people" (DOH 6/23/80)]. There is only the bare assertion of no abnormalities. If questionnaires, blood studies, and physicians' reports were correlated and a report on these provided to the panel, there is no written description of such a systematic study that would have enabled the panel to render an independent judgment on the validity and reliability of the work in order to arrive at their own firm conclusions about acute liver disease, hemolytic anemia, and agranulocytosis.

The panel's report also stated that there were "certainly no peripheral or central nervous system syndromes" (Thomas 10/80). The report's bibliography includes no studies of the central nervous system.

Chronic Health Effects

Neurologist Stephen Barron's study of *chronic* effects on peripheral nerves (mentioned earlier in this chapter) states that, "After controlling for age and gender . . . the two nerves that one would have predicted to show a trend toward slowing if there were neurotoxicity in the 'Love Canal' group did, in fact, show that trend although statistical significance was not reached." (Barron, 1980). Dr. Barron reported that the probabilities that the observations he made in this case had occurred by chance, were .005 and .01.

The level of statistical significance he adopted to "correct" for the fact that fourteen comparisons were made, was .0036, or thirty-six in ten thousand. Thus, even though one of Barron's figures reached the probability level of one in one hundred, the other of fifty in ten thousand, they were not accepted as *statistically significant*, despite the fact that the commonly accepted level is five in one hundred.

Dr. Barron did not do the statistical work alone. He acknowledges the EPA's Division of Neurotoxicology for their work in performing the multivariate regression analysis of his data on May 20 and 21 (Barron 1980). On Thursday, May 22, Dr. Barron testified before a Congressional subcommittee and described his feeling of uncertainty about the interpretation of his data after his two-day session with the expert statisticians. Although he had once believed that two results in his data were statistically significant, and he knew that they might be *clinically important*, at the Congressional hearing he felt that he had to defer to the EPA's statisticians (USHR 5/22/80). Dr. Barron states in his report, "It is noteworthy that the trend of slowing in the 'Love Canal' group is seen at all in view of a protocol that was designed to strongly bias the measurements away from that observation by the rigid exclusion criteria described. . . . Failure to reach statistical significance may have been the result of the number of other measures made, and the slowing may, in fact, have medical significance" (Barron 1980).

The Thomas Panel's dismissal of his findings as "essentially negative" and their unqualified assertion that there are "certainly no" peripheral nervous system syndromes among Love Canal residents should be judged against Dr. Barron's view of the significance of his work. Let us consider further the Thomas Panel's conclusion that "chronic health effects of hazardous waste exposure at Love Canal have neither been established or ruled out yet, in a scientifically rigorous manner" (Thomas 10/80).

The panel discusses three studies of *chronic* health effects in addition to Barron's peripheral nerve conduction study discussed above. One was the chromosome or cytogenetic study. The panel substantiated strong criticisms of the Biogenics Corporation work with this statement, "The Biogenics report was reviewed by several groups of experts in the field of cytogenetics, with expressions of doubt that the reported results were of significance" (Thomas 10/80).

The panel relied on a journalistic attack on Dr. Picciano and his work (Kolata 6/13/80), characterizing it as "an extensive review." That extensive review mentioned only two of the four reviews available at the time it was written. While the Thomas Panel conceded that Dr. Marguery [sic] Shaw's review supported Picciano (Thomas 10/80), the four other reviews had been mailed to Dr. Thomas (Gage 6/30/80). Two were supportive of Picciano's work and two were critical. Dr. Kilian's group, whose report supported Picciano, had visited the Biogenics Laboratory and examined the slides. Dr. Sidney Green sent a favorable report to the EPA following his visit to Picciano's laboratory where he also viewed the slides. Neither of these reviews was mentioned in the Thomas Report.

The third study cited by the Thomas Panel as a report on chronic illness was called *The "Epidemiological Study" by Beverly Paigen, Ph.D.* Thomas, 10/80). The panel minimized her study by casting aspersions on her as a "private citizen" who "believes fervently" in her findings. Further, "The Panel finds the Paigen report literally impossible to interpret. It cannot be taken seriously as a piece of sound epidemiological research, but it does have the impact of a polemic" (Thomas 10/80).

The Panel made some important omissions. Health Commissioner Axelrod, for one, had publicly commended Paigen and the homeowners for their work.

Second, the HEW-EPA committee review, which supported Paigen's findings and was issued publicly in July 1979, was sent to Dr. Thomas on June 30, 1980, by Dr. David Rall, the director of NIEHS who headed the committee and signed its report (Rall 7/2/79, 6/30/80, 12/29/80). The Rall panel review is not mentioned in the Thomas report.

Finally, the Thomas Panel minimized evidence of health effects by dismissing Paigen's study as methodologically inadequate, nowhere specifying its inadequacies. Paigen presented the study as a limited one. However, the only segments of her study examined by the DOH using their own extensive data, supported the differences in miscarriage rates by wet-dry location. Because none of the rest of her suggestions were tested, they remain open as possibilities.

The fourth set of studies of chronic health effects cited by the panel were "the New York Department of Health's epidemiological studies." The panel minimized the statistically significant results of the studies of fetal wastage. "The investigators [Vienna et al., Department of Health] thought there might be some increase in miscarriages and infants with low birth weights, but the data cannot be taken as more than suggestive" (Thomas 10/80).

The study referred to ("Adverse Pregnancy Outcomes in the Love Canal Area") shows rates of miscarriages ranging as high as 50 percent of pregnancies in the mid 1960s and excess rates in wet areas in recent years (Vianna et al., 1980). Vianna et al. reported statistical significance despite the use of control groups that would result in an underestimate of the excess rates of miscarriages. (See Bross 1980)

Again, the Thomas Panel omitted the well-known fact that two New York Health Commissioners had based important policy decisions and recommendations upon the evidence from the DOH data, of increased miscarriage rates and other signs of fetal problems at Lovel Canal.

Included in the panel's materials were some of the correspondence between Dr. Axelrod and Lois Gibbs, including a letter to Lois Gibbs in which Dr. Axelrod wrote,

> I believe we both agree that there is a statistically significant increase in miscarriages and low birth weight babies among families in the area in question, although there remain differences of opinion over methodology. All of our actions to date, including the decision of my predecessor, Dr. Whalen, have been predicated on our belief and that of an expert panel of outside scientists that exposure to chemicals buried at the canal may indeed be responsible for the adverse health effects noted to date (Axelrod 8/21/79).

Obviously, the data were more than suggestive to Commissioners Whalen and Axelrod.

The Thomas report continued its discussion of the DOH's epidemiological studies, "No cases of chloracne were found, and there appeared to be no excess of cancer, asthma, epilepsy, liver disease or hematologic abnormalities" (Thomas 10/80).

The study the panel cited as the source for the statement is the very same study by Vianna et al., "Adverse Pregnancy Outcomes." The annotation of the report in the appendix states, "No instances of chloracne and no excess of cancer, asthma or epilepsy were found among these area residents" (Thomas 10/80). This annotation is totally incorrect in reference to this study of reproductive abnormalities.

A one-and-a-half-page memorandum written by Dr. Vianna to Dr. Greenwald on the day before the panel's third meeting lists seven major *efforts* of Vianna's department. The memo's sole statement regarding chronic conditions was "We found no instances of chloracne and no apparent excesses of asthma and epilepsy among current study area residents" (Vianna 6/19/80). No description of methods, no tables, and no description of a control group appears anywhere.

On page 2, Vianna's memo states that using the Tumor Registry, "Our initial impression was that both the numbers and types of cancer were in no way unusual from that which would be expected in the general population. This initial evaluation is obviously limited." (Vianna 6/19/80)

With no other documentation, and with no way of assessing the methodological adequacy, these summary statements were all that could have provided the basis for the panel's unequivocal conclusions. [Months later a study listed in the Thomas report bibliography as "In progress," was reported in *Science*, based on tumor-registry data. It reported an excess of lung cancers among both men and women living in the Love Canal census tract area, but no other type of cancer in excess of rates elsewhere in the city or state. The authors discounted the lung-cancer findings because they were not clustered at addresses closest to the canal. (Janerich et al. 1980)].

There were two other outstanding omissions. One concerned the episode in the fall of 1979 when some 125 Love Canal families lived in motels at state expense with their acute canal-related illnesses attested to week after week by forty-eight physicians, according to the materials prepared for the panel by the N.Y. Health Planning Commission (N.Y. Health Planning Commission 1980). The second omission concerns the remainder of the massive studies undertaken by the New York health department. While the Thomas Panel characterized these as "aimed primarily at determining reproductive abnormalities," in fact, the questionnaires were far more extensive than that. Two years after studies costing over $3 million were launched by the DOH, two years in which the "State Health department . . . devoted over 205,000 staff-hours (the equivalent of 122 staff years) in a continuing quest to determine the chemical risks . . . [at] Love Canal" (N.Y. DOH 6/23/80), the Thomas Panel might have inquired why there were no studies published or distributed by the DOH. It was in the absence of data that small, pilot projects, undertaken with few or no resources, assumed great importance because they were the only studies publicly available. We might raise those questions, but the Thomas Panel did not.

The Social Context of Evaluation

Although the Thomas report begins by chiding former Commissioner of Health Whalen for the strong language in the brochure, "Love Canal-Public Health Time Bomb," the panel expressed its understanding that "a finding of 'great and imminent peril' was a statutory requirement of Section 1388" to declare the health emergency (Thomas Report 10/80).

This observation may be perfectly sound, but in a context, in which all health research save that of the New York Department of Health is denigrated, and the only criticisms of the New York Department of Health concerned strong language and the lack of coordinated studies (for which both state and federal agencies were chided), the Thomas panel's comments provide a balm and a rationale for the DOH behavior. There are no criticisms of any work that was the responsibility of the health commissioner in office at the time the Thomas panel did its work.

In order to better understand the protective stance adopted by the Thomas Panel towards the New York Department of Health under its current Commissioner, let us consider the social context within which the report was produced. Earlier I described the general political and social tumult during which the panel was appointed by Governor Carey on June 4, 1980. The panel was assigned the task of evaluating medical and scientific data collected at the Love Canal by federal agencies, by "private" (or "other") sources and by the New York Department of Conservation, and by the New York Department of Health. The Commissioner of Health for the State of New York, with responsibilities for the quality of all health care offered within the state, has vast statutory powers to enable him to perform that difficult task. For example, the New York State Health Commissioner must approve every capital expenditure for construction and equipment in all public and private medical facilities (N.Y. Public Health Law, Sec. 2801[5];2802). The Health Commissioner also participates in setting rates for hospital services on an annual basis (Sec. 2807). Although characterized by DOH spokesmen as an "outside and impartial panel" (Shribman and MacClennan 6/5/80), in fact, four of the five physicians assigned to critically evaluate the health department's work were administrators in medical institutions regulated by the Health Commissioner.

Moreover, high-level health-department officials attended all the meetings of the panel and helped shape its final report. The Thomas Panel met on five occasions (Dowling 2/18/80). In addition to the panel members, there were two staff helpers, Dr. Peter Greenwald who identified and obtained pertinent studies (Greenwald 12/3/80) and Edward Dowling who made arrangements for the meeting (Dowling 12/3/80).[a] Their names and affiliations do not appear in the report.

The final meeting was held on July 21, 1980. An initial draft of the report was ready for discussion, but not only by the panel members who

[a]Dr. Greenwald is the director of the Division of Epidemiology for the N.Y. State Department of Health, the division that conducted the state's major health studies at Love Canal, and author of one of the studies listed in the appendix. He attended all but one meeting. Edward Dowling is the associate director of the N.Y. State Health Planning Commission.

signed it. In addition to Greenwald and Dowling, Dr. Kevin Cahill and Jeffrey Sachs, both Governor Carey's aides, attended. On this occasion Mr. Dowling took notes and circulated them to all the participants (Dowling 7/22/80). Of the twenty-eight comments Dowling recorded, twelve are by Governor Carey's special assistant, Dr. Cahill. In other words, the final report was substantially shaped by the contributions of political aides and health department employees, and their contributions to the report and the panel's deliberations were nowhere acknowledged.

The final report, dated October 8, 1980, was submitted to Governor Carey and distributed to all members of the New York State Legislature, to a number of people in federal agencies, and to the press. Reporters started telephoning Beverly Paigen at 4 p.m. Friday, October 10. When Paigen said she could not comment on the report, for she had not seen it, one reporter photocopied it and took it to Paigen. From this copy, Paigen learned that her work was "almost impossible to interpret", and the like. To this day, no copy of the report has been sent officially to Dr. Beverly Paigen, Dr. Stephen Barron, or Dr. Dante Picciano, all of whom suffered from the wide publicity afforded this document.

Contributions to Understanding

When he requested the cooperation of the Love Canal Homeowners Association's president in locating scientific studies for review, Dr. Thomas wrote to Lois Gibbs:

> I am convinced that the Panel's comprehensive evaluation will contribute to the understanding of the public, government agencies, *and the scientific community*, respecting findings to date and request future inquiries [*sic*]. I look forward to sharing the Panel's report with your group. (Thomas 7/3/80, emphasis added)

The promise to share the panel's report with the residents' group was not fulfilled, for no copy of the final report was sent to Gibbs or to any other organized residents' group. The panel's contribution to the scientific community's understanding of the "assessment of health effects from hazardous wastes at the Love Canal" is also virtually nonexistent.

Whether or not they were aware of it, the Thomas panel members had conflicts of interest from the moment they consented to serve on this body, established when the New York governor and health department and other agencies were publicly under attack. They had conflicts, both by commonly accepted standards of ethical behavior and perhaps by legal definition (N.Y. Public

Officer's Law, Sect. 74). The commissioner of the department whose work they were to evaluate had extensive financial powers over their institutions. All meetings were attended by members of the very department whose work they were to evaluate, and some were directly responsible for the work; the final report was shaped in part by the unacknowledged work of members of the health department and the governor's staff. There were no minutes of meetings, almost no records of how work was done, and no record of votes on the final report. The report itself provides no citations to specific documents, draws conclusions on matters of vital importance with no apparent scientific evidence, lists one nonexistent document, omits information, and provides questionable interpretations of data. The panel members and chair refuse to answer questions and will not provide, in a manner customary for scientists, the documents they examined.

The Thomas panel has contributed to the public's distrust of legitimate authority and has raised new questions about the degree to which public health departments are politicized. By its *ad hominum* attacks and sweeping indictments of the work of all agencies and all researchers except those currently in the New York Health Department, the Thomas report may serve to deter nonestablishment and, particularly, younger scientists from undertaking research in this new area of public health concern. Such a consequence would not be totally inadvertent, however. The report's major set of recommendations emphasizes centralization of the administration of studies of Love Canal (as a model for future studies of toxic-waste effects) under the control of the state government. Federal funds should be funneled through one federal agency, which will relate to the state (Thomas 10/80 p. 26). The panel recommends that the studies be supervised closely by a scientific advisory panel of 10 to 15 scientists from all the technical disciplines involved at Love Canal. The special advisory panel would be organizationally housed within the governor's health advisory council (Thomas 10/80).

Although the lack of coordination indeed provided some of the problems at Love Canal, *one* official voice would have had worse results for the Love Canal people and for the larger public. There are ways to avoid the combined chaos and neglect that the panel deplores while still maintaining the crucial scientific control functions that Merton has called "organized skepticism" (Merton 1973). Campbell, for example, suggested the use of simultaneous replication of studies, making data readily available for reanalysis, careful attention to minority reports, and protection for "whistle blowers," as approaches to the similar political and scientific problems presented in evaluating large-scale social programs (Campbell 1979).

The Thomas panel's recommendations for centralized studies, with an advisory panel virtually in control, are based on four untenable assumptions: (1) that such a group indeed knows *the* correct approach or approaches; (2) that other approaches are not possible; (3) that there could be no differences in interpretations of the outcomes of studies; and (4) that the advisory panel members are to be trusted *because* they are distinguished and eminent. Even the distinguished and eminent, however, need the balance provided by the broader scientific community.

In short, rather than encouraging the scientific norms of independent effort, mutual criticism, and healthy controversy, the Thomas panel urged *one* voice, and only one, of government-sponsored science and recommended that scientific uncertainties not be shared with the public—all in the name of preventing the public from worry, anxiety, and undue concern about the public's health! Public concern and public outcries when tax-supported health agencies are not doing what they should do for the people apparently create political problems—problems that surely would be solved by a reassuring, noncontradicted voice of government-supported "science."

As Goldsen has pointed out in discussing the importance of information in a democratic society:

> Governments can govern only if people are governable. . . . What makes them so is a state of mind. . . . When we feel that it is all in the name of law, order, morality, tradition, right, justice—when we feel this way in our bones—everything seems right and natural. Commands are virtually unnecessary. (Goldsen 1977, p. 275)

"The Miracle of Love Canal"

In response to the scientific controversy, and the Thomas Panel's emphasis on psychological damage, Love Canal resident Marie Pozniak said:

> Gee! It's the miracle of the Love Canal! You live on top of a couple of hundred chemicals for 25 years and nothing's wrong! You just *think* something's wrong! They should really send all the world's wastes here, because it's the only place on earth such a miracle could happen.

The battle described in this chapter was a gloves-off, no-holds-barred struggle. Those involved used the power of government and access to the press to write their own versions of what happened at Love Canal. From the very beginning, the definitions of the Love Canal health and environmental problems—where they are, what they are, how serious they are—have varied considerably, depending on who defined the problem, when, to whom, and what they stood to gain or lose from the definition. The words, however, were usually spoken in the language of science.

At present, there seems to be a semiofficial contemporary legend emerging about Love Canal, composed of the following elements:

1. There are no proven physical health effects at Love Canal.
2. There is only psychological damage to the people.
3. The psychological damage was created by
   a. fumbling bureaucrats,
   b. overzealous scientists,
   c. the mass media,
   d. corporate enemies,
   e. screaming housewives
   f. rabid environmentalists,
   g. any or all of the above.
4. People were moved out of the Love Canal area only for compassionate reasons—the psychological damage—or for practical reasons—to allow construction work to proceed in or near their property.

A derivative of this basic tale is that, if any toxic-waste public-health research is to be done, only health researchers should do it, with no interference. Two underlying premises are (1) that Love Canal is unique, not the tip of an iceberg;[13] and (2) that chemicals are innocent until proved guilty beyond the shadow of a doubt.

The reason both for the contest and for the emerging explanation is very clear. The stakes are huge—involving not only careers and reputations but vast amounts of money. What powerful interests do this explanation and its variants serve best?

1. The state of New York has spent some $61 million in direct, easily measured costs at Love Canal, and the city of Niagara Falls about $7 million. The federal government, according to state estimates, will reimburse the city and state for $21 million of that amount (NYDOT 12/31/80; NYDOH 10/81). The people responsible for allocating and spending those resources want to see the problem solved and to believe, and have others believe, that they did the right thing and that they *solved the problem*—for the city, for the state, for the country. They want to emphasize that what was done was necessary and that what was necessary was done.

2. In 1979, the federal expenditure for environmental health research was $622.8 million. For fiscal year 1981, the NIEHS estimates that $728 million will be spent. (NIEHS 7/80). Who will control these funds? It will surely not be agencies whose multimillion-dollar projects are thrown into doubt by people working with small amounts of money or no money at all.

3. The Hooker Chemical Corporation is the defendant in lawsuits adding up to a few billion dollars. The suits have been filed by the state of New York and by the U.S. Department of Justice on behalf of EPA. A lack of proof of physical harm to persons would be very helpful to a company facing these and hundreds of personal-damage lawsuits.

The Hooker Corporation widely disseminates the pronouncements of scientists who even express doubt that there are chemical-related health effects at Love Canal, let alone unequivocally stated positions. The vast amount of public-relations effort shows how important those statements are to a company that is doing what it considers proper for its interests.

4. The question of the regulation of the disposal of toxic wastes has been the subject of congressional consideration for several years. In the autumn of 1980, after two years of congressional hearings, a regulatory bill, the "Superfund" (P.L. 96-150, 96th Cong., 2d Sess.), was passed, with an important political compromise built into it. There are no provisions in the Superfund legislation to compensate people for health damage.

The intricacies of implementation of this bill will now proceed under the leadership of a political regime that is unsympathetic to the regulation of industry in general. We may speculate whether "no evidence of harm" will be neatly transformed into "no harm" when an industry worth $113 billion—some 7 percent of the gross national product (U.S. Department of Commerce 1978)—finds that particular data interpretation the more correct one. The chemicals are innocent until proved guilty—an appealing assumption—protective of the chemicals' "civil rights" and statistically supportable as well. While the "trial" takes place, with the full panoply of due-process rights for the chemicals, the less expensive means of chemical-waste disposal can continue.

The emerging legend about Love Canal and what happened there promotes powerful interests and only hurts or might hurt apparently unimportant ones: the annoying finger pointers; those who inadvertently stumble into protected scientific turf; the people of Love Canal, who do not know what has happened to their health; and the people of the United States, who, if they accept this version of "the miracle of Love Canal," will be protected from worry about the long-term effects of toxic chemicals until, perhaps, it is too late to do anything about them.

The Love Canal disaster is as much a symbolic, conceptual event as it is a physical one. At Love Canal, we have seen how various contenders strove to have their versions of the truth generally accepted. At Love Canal, we have been witnessing the struggle for the final history that will remain after the events are over—to justify and to guide decisions not only about this Love Canal but about future Love Canals as well.

Notes

1. In addition to the suit pertaining to Love Canal, the three others concerned other disposal sites in Niagara Falls: Hyde Park, 102nd Street and the "S" dump area. The Hooker Chemicals and Plastics Corporation,

Hooker Chemical Corporation, Occidental Petroleum Investment Corporation, and Occidental Petroleum Corporation were named as defendants in all four. The Olin Corporation was also named in the 102nd Street suit, and the city of Niagara Falls was also named in the "S" area suit.

Only the Love Canal suit identifies the school board and the county health department as defendants, along with the city and the corporate defendants (*USA* v. *Hooker* 12/20/79). In that suit, the EPA charged Hooker with disposing of over 21,000 tons of various chemical wastes in Love Canal in such a manner that the escape of hazardous wastes presented an immediate and substantial endangerment to the health and the environment in violation of the Resource Conservation and Recovery Act (Sect. 7003), the Clean Water Act (Sects. 504 and 309, 33 U.S.C. 1364), the Safe Drinking Water Act (42 U.S.C. 300i), and the Rivers and Harbors Act (33 U.S.C. 407). Furthermore, the EPA charged that Hooker's practices constituted a public nuisance. The EPA asked for injunctive relief, restitution, and penalties.

On May 19, 1980, the Hooker Corporation filed denials and affirmative defenses; on October 28, 1980, Hooker filed cross claims against the city of Niagara Falls and the Niagara Falls Board of Education and counterclaims against the state of New York and the United States, as third parties.

2. The "impacted area" was defined as 93rd Street to 103rd Street, Cayuga Drive to Buffalo Avenue. This area was larger than that earlier designated as the Love Canal neighborhood. The new designation was based on the area designated a year previously for tax relief by the state legislature, in an attempt to stabilize the neighborhood (N.Y. Real Property Tax Law 1979; see Komorowski 12/21/78).

3. My clipping file of articles from the *Buffalo Evening News, Courier Express, Niagara Gazette* and the *New York Times* contains 97 articles on Love Canal for the month from April 17 to May 16, 1980, and 122 articles for the month from May 17 to June 16, 1979, compared with 544 articles between May 17 and June 16, 1980. In addition to radio and national television news spots day after day, a number of national television shows during that period were devoted partly or wholly to the Love Canal problem, including "Today" (5/22/80), "The MacNeil-Lehrer Report" (5/22/80), "Sixty Minutes" (5/25/80), and "CBS Sunday News" (5/25/80). The "Phil Donahue Show" based an entire hour on the problem, busing forty Love Canal residents to Chicago for the taping, and bringing Lois Gibbs and Mayor O'Laughlin by airplane (June 19, 1980). "Good Morning America" devoted several sections to Love Canal over a two-week period in early June. There were documentaries as well, including one by PBS featuring Luella Kenny whose seven-year-old son died in the autumn of 1978 (7/25/80), a half-hour of a two-hour "Nova" show—"A Plague on Our Children", an updating on CBS of "The Killing Grounds," which had a

fifteen-minute segment on Love Canal (8/12/80); and a documentary broadcast by the Westinghouse stations.

4. Dr. Carter has not responded to my letter asking him whether the events recounted by Dr. Picciano are factually correct and whether he would like to add anything to the account (Levine 3/31/81).

5. The report was signed by Dr. Michael Bender, Brookhaven National Laboratory; Dr. Arthur Bloom, Columbia University; and Dr. Sheldon Wolff, University of California at San Francisco. The cover letter submitting the report to the USEPA was signed by Dr. David Rall, director of the National Institute of Environmental Health Sciences. Because he empaneled the committee, the review is referred to as the Rall report.

6. A sixth review was performed in addition to the five submitted to EPA. Dr. Marvin Legator, Department of Toxicology, School of Medicine, University of Texas, visited the Biogenics Corporation laboratory early in June 1980 and examined the slides and photographs used in the pilot study. Dr. Legator prepared his report as a consultant for the Hooker Chemical Corporation, and the report is therefore not publicly available. Dr. Legator told me, however, that the methods used in preparation and analysis of the slides were as good as the state of the art allows, and he agreed with Dr. Picciano's findings and conclusions. Although he also agreed with Dr. Picciano that the design of the study was flawed by the lack of a comparison group, Legator said that the execution was exemplary and that Dr. Picciano had stayed well within the appropriate bounds of inferences that could be drawn from the materials he worked with (Legator 4/16/81; Levine 4/17/81).

7. In addition to Dr. Kilian, the group consisted of Dr. C.B. Jacobson, associate professor of genetics, George Washington University, and Dr. R.G. McKinnell, professor of genetics and cell biology, University of Minnesota, St. Paul. Dr. Picciano had worked and published with Dr. Kilian in the past, but Kilian had had no part in the EPA-sponsored study of the Love Canal residents.

8. The members of the committee chaired by Dr. Albert were Dr. Arleen D. Auerbach, Memorial Sloan-Kettering Cancer Center; Dr. Maimon Cohen, Northwestern University; Dr. Kurt Hirschhorn, Mt. Sinai Medical Center; Dr. Harold Klinger, Albert Einstein College of Medicine; Dr. Peter Nowell, University of Pennsylvania; and Dr. Neil Wald, University of Pittsburgh. Drs. Sidney Green and Peter Voytek of the EPA served as observers.

9. In March 1981, Dr. Picciano filed suit against *Science* claiming that he was libeled in the article (see Warner 4/1/81).

10. The Love Canal people were interested in health studies, however. During the very period that the Love Canal residents refused to participate in government-sponsored health studies, several hundred began to participate in a study of children's growth and development, conducted by

Dr. Beverly Paigen and Dr. Ted Steegman of the Department of Anthropology, SUNY/Buffalo (see *Buffalo Courier-Express* 8/12/80). The results of that study are not yet available.

11. The Moreland Act is embodied in New York's Executive Law, Section 6. The executive is empowered to examine and inspect the operation of state agencies, with powers of subpoena and interrogation of witnesses under oath. For a description of a prolonged Moreland Act hearing into the operations of New York State's Urban Development Corporation, see Osborn (1977).

12. Mr. Edward Dowling, Jr., associate director of the New York State Health Planning Commission, served as a staff person for the Thomas Panel. He supplied me with materials and lists of materials that he had made available to the panel, as well as some procedural information, in response to my requests made pursuant to New York State's Open Meetings Law and Freedom of Information Law (N.Y. Public Officer's Law, Sec. 101 and 87, respectively). The panel's secretary, Dr. Saul Farber, has never responded to inquiries, nor has Dr. Thomas responded beyond his letter stating that he would not respond to any questions.

13. The "uniqueness of Love Canal" position was expressed well by Charles L. Sercu of Dow Chemical Company—chairman of the Environmental Management Committee of the Chemical Manufacturers Association—in a formal presentation, press conference, and responses to audience questions at the annual meeting of the American Association for the Advancement of Science (AAAS) in Toronto on January 7, 1981. Robert F. Kelly, a Union Carbide Corporation representative, expressed the same views and cited the Thomas report as showing chemicals' harmless nature in a speech to a meeting of the Society for Occupational and Environmental Health in Washington, D.C., December 9, 1980 (see Moe 12/14/80).

7 Help and Self-Help

The Love Canal residents began to organize very soon after their situation was publicized in the spring of 1978. Eventually, there were several citizens' groups at Love Canal, all serving as political pressure groups to some degree, and all serving their members as self-help supportive groups. The focus in the following discussion will be on the Love Canal Homeowners Association, for it was the major group supporting the residents' struggle at Love Canal. Founded and led by residents, it was the first formally organized group on the scene, and it was also first in size, staying power, accomplishments, and local and national visibility. Other organizations will be described as part of the dynamic social environment within which the Homeowners Association thrived.

By the evening of August 4, 1978, when some six hundred Love Canal residents met to organize the Love Canal Homeowners Association, they were no longer the people they had been a few months earlier—chiefly concerned with their private lives. They had spent the spring and summer learning about long-buried chemicals that were poisoning their neighborhood, their children's elementary school, their properties, their homes, their bodies—almost everything in their lives.

They had turned to each other more and more as the summer dragged on, discussing and assimilating the news that a series of corporate and governmental decisions made long ago, combined with the persistent lack of regulations, were now victimizing them. There had been sporadic talk of organizing for some time. A handful of residents had spoken about a tax-and-mortgage action. Lois Gibbs and Debbie Cerrillo had trudged from door to door and collected names to support a citizens' committee to address the Love Canal problem.

When Health Commissioner Whalen issued his order on August 2, 1978, the spark was lit that inflamed an already heated community. By declaring that the hidden dump site in their midst posed "an extremely serious threat and danger to the health, safety and welfare of those . . . living near it" (Whalen 8/2/78); by issuing an order with recommendations that some people move temporarily, but as soon as possible; by failing to mention help for such moves; and by issuing the order in a city hundreds of miles from the Love Canal area, the health department inadvertently proved to the residents that they had to organize to protect themselves. The people joined together and settled in for what they sensed would be a long fight—one they thought might take weeks or possibly months to win. It would be more than two years, however, before some of their goals were accomplished.

Shared Beliefs: What They Were and Why They Developed

On that hot night of August 4, 1978, when the people formally established the Love Canal Homeowners Association, they already shared some beliefs and explanations about the puzzling and frightening turn their lives had taken since the early spring. It did not matter that they did not yet know the precise names of the chemicals and their locations, nor the precise causes and effects of environmental and health damage. Those who gathered in the local firehall and cheered as they elected Lois Gibbs to represent them certainly knew that they had problems they had never had before. Had they stated their shared beliefs about the problems, they would have included the following elements:

Belief 1: *We are the blameless victims of a disaster.*[1] The Hooker Chemical Company, and perhaps others, put huge amounts of chemical wastes into the ground. They, not we, are to blame. Various government agencies, responsible for our health, safety and welfare, failed to care for our interests properly from the very beginning, in 1953, until the present moment. They, not we, are to blame. The state, in trying to do something about the problem, created further problems for us by declaring that the area was not safe to live in. They, not we, are to blame.

Belief 2: *The problems we face are too large for us. We need help.*

Belief 3: *We are good citizens. We deserve help from the government.* We pay taxes and are patriotic and law-abiding. We do our part.

Belief 4: *The government can and should help us now.* One woman summed it up:

> We were a proud community. We are workers and we've worked hard and paid our bills and our taxes, and we raised our kids and didn't ask for a handout. And now we don't want handouts, we just want to be treated fair. This thing isn't our fault, and it's just too big to handle.

Additional beliefs developed in reaction to events. The residents' first victory came quite easily. A few days after the health commissioner's order of August 2, 1978, with its modest recommendations (pregnant women and families with children under age two should move temporarily), Governor Carey said that the 239 families living next to the canal would be moved out by the state, if they chose, and on generous terms (Perrault 8/8/78).[2]

Over the next few weeks, there were some further concessions to the organized residents, who now focused on the problems of outer-ring residents. At their leaders' insistence, the Homeowners Association was consulted about a safety plan to protect residents during the proposed construction project; the organization was given a large office in the 99th Street School; the leaders were allowed to attend some task force meetings; and a toxicologist was hired as a consultant to the association at the state's expense.

However, the governor's assurances that there would be further considerations about moving people once the health data were collected among outer-ring residents came to naught. By mid-September, the outer-ring residents added this important idea to their beliefs about their situation:

Belief 5: *We are being treated unfairly.* The inner-ring people were in little more danger than we are; no one knows how much danger we are all in. Because we are no different from them, we deserve to get exactly what they got, but we are not getting it.[3]

Belief 6: *We must stick together to take care of ourselves.* From the late spring and early summer of 1978, some Love Canal people had thought that perhaps they *ought* to join forces. In the heat of their angry reaction to the health commissioner's order of August 2, 1978, they did join together. For many residents the feeling grew that the state was basically indifferent, or worse, to the needs of the residents who were "left behind." The newspapers were filled with stories of the migration of chemicals and the discovery of dioxin, coupled with health department denials and cautious statements about health effects. The residents' telephones rang repeatedly with DOH requests for further information on the health surveys already completed and often with the news that blood samples must be retaken. Meanwhile, the residents were not getting what they considered enough straight information from the authorities. They were rapidly losing their trust in the government officials and scientists. In addition, their earlier victories showed them that their concerted efforts could produce results. "We *ought* to stick together" became "We *must* stick together."

Belief 7: *Family and community help is not enough for our needs.*

Belief 8: *No one but the government has enough resources for our pressing needs.*

Belief 9: *We must work together to force the government to provide us what we are entitled to.*

As time went on, beliefs about their need for help, about the importance of the government as a prime source of that help, and about relying on each other became firmer.

Other Sources of Help

The outer-ring residents, in particular, found that, even when other sources of help were available, they were insufficient for their needs. We shall briefly examine the Love Canal residents' experiences with such sources of help as their families, the local community, relief agencies, the local university, and the churches, and suggest reasons for their inability or unwillingness to help the Love Canal residents to the extent the people thought necessary.

Relatives as Helpers

In August 1978, when television screens and newspapers spread the message about Love Canal, many Love Canal residents heard from their close relatives. Many received sympathetic expressions of concern, and some were visited by worried family members bent on immediate rescue operations. Money, loans of cars, trucks, and living space, help in packing and moving, and child care came from families. Most of the help was appreciated keenly; sometimes, however, the help offers led to other complications ("Imagine the life for all of us when we moved, with two kids under age three, into my in-laws' four room apartment!"). It soon became obvious that help from families, even if generous and graciously tendered—as it usually was—was but a small part of what was essential to truly solve the problems.

Some people, even among the inner-ring group, were not offered tangible help from families. Some families were related to others in the Love Canal area, and, although it gave them a common bond, it also strained family resources. The outer-ring residents found, often to their disappointment, that, although some of their relatives sympathized with their plight, others could not seem to comprehend this "invisible disaster," let alone sympathize with the residents' passionate desire to get out of the Love Canal situation. In any event, well-intentioned or not, families simply did not and could not provide the resources needed by the people at Love Canal.

Local Community

The local community gathered around to some extent when the Love Canal emergency was publicized. Contributions were made to a fund started by the local newspaper, the *Niagara Gazette*, and seeded by its publisher, the Gannett Company (Niagara Gazette 8/6/78). A local supermarket contributed food and food certificates to help replace food discarded from inner-ring home freezers and pantry shelves. The Flickinger Company, a wholesale supplier, donated produce and bakery products. Some stores and moving companies provided cartons to pack household goods. However, once the inner-ring residents' immediate problem seemed on its way to solution, and a massive construction project was under way to cover the canal, there were few concerted offers of community help to the outer-ring residents.

There were some outstanding exceptions. Ralph Kushner, the owner-operator of two Niagara Falls supermarkets, donated $3,000 worth of food to the inner-ring residents and later held fund-raising events ("supermarket sweeps"), netting the Homeowners Association some $5,000 between the fall of 1978 and 1980. Kushner personally donated $1,000 to the treasury of the association when it was at a low point. He also was active in coordinating a group of merchants who contributed $3,000, which went partly to the Love Canal Homeowners Association for office supplies and partly to the United Way to help Love Canal residents (*Niagara Gazette* 8/13/78).

Starting in the spring of 1979, local branches of the United Auto Workers, the Oil, Chemical, and Atomic Workers Union, the International Association of Machinists, the Teamsters, WKBW (the local ABC television station), and other employee groups donated a total of a few thousand dollars. The New York State United Teachers and the Buffalo Teachers Union passed resolutions of support for the outer-ring residents months later.

Offers of sympathy and help—and the help itself—were much appreciated by the residents. It was soon obvious, however, that the community could not take care of its own, because the costs at Love Canal were going to run into the millions—outstripping the resources the local community could contribute. Furthermore, the nature of the Love Canal disaster had an effect on the community's motivation to contribute. There were no walls of water, no bolts of lightning, no reports of multiple deaths and brave rescues. In short, the Love Canal situation was neither cataclysmic nor dramatic, and, as a result, there was no mass outpouring of help and concern from the various business, civic, and other private organizations that is common in times of natural disasters. In addition, once the state purchased the inner-ring homes, many local people assumed that the worst problems had been taken care of, and they did not know about, understand, or perhaps care about the plight of the outer-ring people.

Relief Agencies

During the first weeks after the announcements that people *should* move and then that inner-ring residents *would* move, the Red Cross and Catholic

Charities provided help to people in the form, for example, of temporary loans, help with moving, and vouchers for new mattresses. The Red Cross provided some crisis counseling as well. It is significant that the forms of help provided were familiar ones for these organizations. Large service organizations can do best what they have been doing; difficulties arise when the situation calls for new forms of help (see Dynes and Quarantelli 1970; Dynes and Aguirre 1979; Stallings 1978; see also *Buffalo Courier-Expess* 8/12/78).

The United Way of Niagara provides our first example. From the first days of the officially declared emergency at Love Canal, the United Way offered help repeatedly, through newspaper announcements and brochures, and provided many Love Canal people with the services available from its twenty-three private agencies (*Niagara Gazette* 8/12/78). Day care for children was particularly appreciated by parents who were abruptly faced with a search for new living quarters and a multitude of extra tasks. The United Way served as a clearing house for the contributions raised by the merchants and other groups and also arranged for housing three outer-ring families in motels for almost two months (*Niagara Gazette* 10/6/78).

In December 1978, however, the United Way was called upon to do something new, when William Hennessy, chairman of the governor's interagency task force, announced that the state of New York would allocate $200,000 to the United Way of Niagara to provide residents with some new assistance for medical examinations, blood tests, and "other services" (MacClennan 12/17/78). (As early as August 31, 1978, the Homeowners Association had asked for extra funds to cover the costs of repeated blood tests and other medical expenses.) Hennessy's announcement was part of a financial package (including a possible tax reduction) that apparently was intended to mollify the Love Canal residents, who were attracting media attention at that time by picketing the massive remedial construction work at the canal. By the end of March 1979, the United Way proposal had been approved for spending the funds on health-related services in four categories: medical, mental health, recreational, and informational. The proposal emphasized the very services United Way agencies were already providing (NYDOT Spring 1979), with the exception of the medical category.

In the long run, although $75,000 was originally budgeted for medical expenses, only $20,000 of the funds was spent for medical purposes (Maloney 7/7/81). By the summer of 1981, the United Way executive director thought that, since most people have medical insurance, they simply had not applied for extra funds for medical purposes (Maloney 7/7/81). There were different views, from the beginning, about what the funds could be spent on, however. The United Way proposal stated that certain blood tests and a physical exam would be reimbursible. "Additionally, other tests and services would be ruled eligible upon the recommendation of the State Health Department and the local Medical Society" (NYDOT Spring 1979, p. 3). The eligible tests were described to residents as diagnostic. Repetition of tests already done by the department of health were reimbursable, but other meanings of *diagnostic* varied.

On October 21, 1979, for example, Lois Gibbs requested funding from the United Way special Love Canal grant "to study and evaluate the effect of living in the Love Canal area on the male reproductive system" (Gibbs 10/21/79). The amount requested was under $10,000—to cover the costs of medical examinations and blood and semen analyses for twenty to thirty men at about $320 each. In his letter to Gibbs refusing the request, Joseph Maloney, explained that, after her request was submitted to "various State and local Medical people," it was disapproved on the grounds that the state's grant to the United Way was provided for diagnostic testing, not for research (Maloney 11/9/79).

The extra funds allotted by the state to the United Way for Love Canal assistance were chiefly devoted to camping and day care for the Love Canal neighborhood children. By September 1979, $90,000 of the $200,000 had been spent, according to United Way's accounting: some $53,000 on day care, camping, and field trips; $16,000 on salaries; and almost $12,000 on United Way overhead costs. Only $3,800 was spent on medical bills and $32.54 on direct mental health services. Almost all the remaining $5,000 (of the $90,000) went toward a conference for professionals and for supplies (United Way 9/79).

Only the first few months' figures were available in July 1981, for the final accounting of the expenditure of the $200,000 was still in preparation. Mr. Maloney estimated that $20,000 had been spent on medical expenses and about $35,000 on a counseling program developed after July 1979. The counselors worked long hours when the people lived in motels in the autumn of 1979, but they remained neutral on the need for relocation of residents, according to the executive director (Maloney 7/7/81).

Although the United Way administrators were aware of the need for flexibility in expending the $200,000, there was a major hindrance to their flexible interpretations of guidelines—essential in a new type of disaster. Organizations must be supported. The United Way, dependent for its support on the goodwill of powerful community forces, is well known for avoidance of controversial stances in relation to government and businesses (Graziano 1969). The role of United Way is to help people, as the executive director said, and not to take sides.

For the residents, the United Way's carefully neutral stance resulted in help that was appreciated but was clearly insufficient for people who were fighting to change a government policy decision. For some, the fact that the agency took the state's money put an aura of suspicion over the agency's actions, no matter how well-intentioned and hard-working they were.

The University Task Force

In the summer of 1978, Wayne Hadley, Lois Gibbs's brother-in-law was a member of the Biology Department at the State University of New York at Buffalo. He enlisted the interest and help of his colleague Charles Ebert, a professor of geography, and together they approached the university president, Dr. Robert Ketter, with a request that he make the administrative arrangements necessary for expert faculty to aid the Love Canal residents. The president, an engineer, had frequently stressed the importance of the tax-supported university's responding sensitively to local community needs, and he recognized the seriousness of many of the Love Canal problems. He consulted with two university vice-presidents and two deans. With their advice, he decided that it was important that the university refrain from rushing into the Love Canal situation, interfering with investigations being done by the state, taking on any sort of investigatory role, or getting involved in the controversies of forthcoming litigation. However, the university president did want to encourage individual faculty members who might be interested in helping at Love Canal (Ketter 1/8/79). Therefore, in mid-August, he appointed a task force (*Buffalo Courier Express* 8/19/78). Concerned that the committee remain and be seen as neutral and objective in its approach, he appointed, as chairman, the dean of the School of Engineering, whose neutrality was evidenced by the fact that he had displayed little interest in Love Canal up to that point.

At about that time, the university president was also asked by the Erie County Executive to find faculty members interested in the general problem of the disposal of toxic wastes. (The University is in Erie County; Love Canal is in adjoining Niagara County.) When the president described the SUNY/Buffalo task force in his annual state-of-the-university message (*U/B Alumni News* 9/78), the newborn task force was already moribund as far as any specific aid to Love Canal residents went, for the defined task was one in which the university would examine toxic-waste problems in Western New York in general. The university task force's ensuing activities consisted of a few meetings, at which the attention of interested faculty cooled because they were given little information and almost no direction by the task force chairman. The headlined promise: "Professors to Study Love Canal" (Delmonte 8/23/78) went unfulfilled. The task force's major tangible accomplishment was the creation of a limited collection of some news articles and reports on Love Canal. The intangible consequences of the very existence of the SUNY/Buffalo task force, however, was that other interested faculty members believed that a university group existed and was doing something at Love Canal. Disappointed Love Canal residents, who had read, with some relief, "UB President Creates Unit on Love Canal" (*Buffalo Evening News* 8/21/78), soon referred to the university group derisively as "the U.B. Task Farce."

The university's involvement must be understood at two levels: the professors employed by the university and the university as an official body. University professors have obligations to the university, but they have great freedom in devoting their efforts to self-selected research and scholarly problems. Thus, Hadley, Ebert, and I participated in the Love Canal situation with no interference from the university's administration—and, in my case, with some official encouragement. The university as an entity, however, is represented through the president and his official designees to the outside world. The difficulties and complications for the state university of working uninvited by state agencies appointed to the Love Canal task force or by the governor who appointed them may have been rather formidable to contemplate.

In this case, the university did two things: (1) appointed a committee and hoped that, in time, hot situations would cool; and (2) it did what it is organized to do best—allowing those who were interested to undertake research when and where they could. The university as an organization, however, could not, or would not, take on the task of providing information and advice so sorely needed by the Love Canal residents.

The Churches

Local religious organizations, in adapting to the Love Canal disaster, eventually played an interesting role. There were three stages in their involvement. During the first two stages, from the spring of 1978 to the late summer and early autumn of 1979, the organized religious bodies went from months of little visible involvement to verbal commitments and preparation to help the Love Canal people. In the third stage, when the organized churches did provide help, it was done through a professionally led organization, The Ecumenical Task Force, which provided part of the dynamic social environment for the Homeowners Association.

Outside Help Is Not Enough

For a variety of reasons, then, during the first Love Canal year, the Love Canal residents felt unable to derive the full help they needed from the idealized "personal and social networks of families, neighbors, and community organizations to which people naturally turn as they cope with their problems" (President's Commission 1978, p. 14).

They knew that any financial help that might be forthcoming as a consequence of lawsuits filed against the Hooker company would come in the future, if at all. Whether because their needs were unrecognized or too large, had to be met too quickly, or were viewed as inappropriate for existing organizations, the residents felt more and more strongly that they must work together to find help, and that the government was the chief source of sufficient and available resources.[4]

In addition to shared beliefs that resulted from the Love Canal residents' views of the causes of their problems, the help they felt entitled to, and the inadequacies of private sources of material help, their feelings of dependence on each other were strengthened by emotional needs that increased as time went on.

Belief 10: *We are the only ones who can understand each other.*

Consider some of the events, feelings, and behaviors the Love Canal residents shared. All, or almost all, experienced feelings of uncertainty about what had happened and what would happen to their health. Parents were unsure of what to do to protect their children. In many cases, one spouse would argue for leaving the area immediately in order to protect the children's health, while the other would argue that, if they did, the family would face certain financial ruin.

People experienced fear when the very word *home* lost its deep connotation of a place of refuge and security. Many people felt fear and helplessness all the time.

People were by no means eager simply to flee their homes, even though they believed the homes were contaminated. Many men had held extra jobs to acquire the down payments and to meet mortgage payments on their homes. Many had contributed their sweat, energy, skills, and time—and sometimes the bartered skills of neighbors—to improve their homes. People spoke lovingly of details of their homes—the picket fences, the pretty yards—that they did not want to leave.

People expressed their sense of the loss of their feelings about their homes as the center of their family life in many ways. For several months, some inner-ring residents returned to visit their homes regularly. Weeping, one woman who was moving out told me:

> My son will be leaving for his wedding, and it won't be from the home he grew up in. He won't dress in his own room. He's going to leave from some strange place.

Outer ring residents—accustomed to viewing their homes almost as living entities as well as growing investments—felt the loss of their interest in their homes, felt their former love of home turning to hatred.

> You work for a home. Now in a few years I'll go to retire and I've got a dead box. I was going to paint up and make a garage and fix up. Now I'm not going to put money in a dead box. . . . I have no desire now to take care of this place.

Most people experienced severe disruptions of their daily routines. Organization activities, conversations, reading, and thinking about Love Canal absorbed their energies. Families who moved in with relatives or to motels exhausted themselves trying to carry on some semblance of their normal routines in unfamiliar settings—some of them unsuitable for daily family living. Some families sorely missed visits from friends and relatives, who were afraid to come into the area. People missed their customary work, hobbies, and recreation—the familiar flow and setting of family life. They missed their neighbors and neighborhood acquaintances—the comforting sense of familiar faces in familiar places.[5]

While children suffered, their parents felt at fault for their childrens' troubles—no matter how blameless they felt at the same time. Young children were moved to new schools when the 99th Street School closed, and young and older students alike often were singled out for unwelcome attention at school, through interested questioning or, very often, teasing by their peers. Children talked about chemicals—the younger ones not certain of what they were but knowing that they were so powerful that even their parents cried sometimes when they talked about them and could not protect the children from their unknown dangers.

The disruptions and uncertainties were intertwined with, and caused changes in relationships, in feelings, and in outlook. Perhaps most important, men felt unable to care for their families by providing a proper home, and parents felt unable to protect their children. Many felt trapped. In the words of one young father, "We've lost all of our options. Can't rent, can't sell, can't walk away from it."

Banding together with other residents not only helped many people feel that they were asserting some control and reducing uncertainty by doing something positive to correct their unwanted, unasked-for condition, it also made them even more dependent on each other as their problems and activities became widely known. Although they attracted attention and many expressions of sympathy, the residents, particularly those from the outer ring, also reported that they were sometimes ridiculed, told that they had been fools to buy homes at Love Canal, suspected of trying to "make a bundle" from the government,[6] accused of giving the city a bad name, sneered at for seeking publicity for its own sake, and feared as the contaminated carriers of mysterious diseases. The comments and reactions stemmed not only from strangers and fellow workers but often from friends and even from relatives.

As time passed and discouraging incidents piled one on the other, residents felt more and more that no one but people undergoing the experience themselves (or at least a similar experience) could understand them. Only fellow sufferers could share the feelings of uncertainty about their health, their children's futures, their feelings of frustration in dealing with the government agencies, and their despair as the months passed and their problems remained unsolved. As they confided in each other more and more, both privately and in public; as they worked together and participated in individual and group actions—particularly actions that were surprising to them even as they did them; as they interpreted their beliefs, their goals, their behavior; and as they thought and planned together for a better and happier future, free of Love Canal, they became essential supports for each other—a sort of large self-help group, a substitute for a strong, supportive, and ever-understanding family.[7] In their words: "We are the only ones who can really understand each other. We can hardly talk to anyone else about how we really feel."

Shared Beliefs: The Purpose They Served

The residents' shared beliefs were intensified and elaborated as time went on. They provided an explanation to the residents about the causes and cures for their problems and attributed both causes and cures to elements in the social structure. The beliefs provided a framework from which actions could logically flow to restore to the affected people some sense of control over their lives.[8] The beliefs were broadly understood and could be easily stated and elaborated to anyone outside Love Canal. The beliefs also provided a set of assumptions that helped weld the people who shared them into a tight, working group and sustained them during the long period of their struggle.

Homeowners Association Resources

The tangible resources of the Homeowners Association are simple to describe. Money was always in short supply. There were some donations, some fund-raising activities, and membership fees of one dollar per family—all resulting in a treasury totaling about $5,500 during the associations first ten months and about $12,000 over a thirty-month period. That treasury barely covered the expenses for the organization's activities.

The state provided an office for the Homeowners Association, for more than two years, paying utility and telephone bills and some copying expenses. This support was critical. The office became a focus for the organization, a place to be, the center of work, the haven for many troubled people, and the place where information was available and where people could share their most pressing concerns. The association president, vice-president, and office workers provided an essential presence in the office. They were there every day and available around the clock by telephone for those who needed them. They provided a stream of newsletters for members, with accounts of the association's, the state's, and all other activities related to their cause. Residents came to the office or called for news, sympathy, warmth, comfort, and just to regain some feeling of normality in a suddenly crazy world.

The Love Canal Homeowners Association's chief resource was the will of a handful of dedicated people—obsessed with the hope of getting out, of obtaining help, and of returning everybody and everything back to normal. That group worked, doggedly, without pay, days, nights, and weekends for almost two and a half years. Approximately ten people or less were very active daily. Another thirty could be called upon at short notice, and many more attended public meetings.

The core group was chiefly women, who somehow managed to take care of their homes, shop, make meals, supervise their children, and still put in long hours working on association activities. There were also some men involved, particularly those who worked night shifts, were retired, were temporarily unemployed, or otherwise found time to devote to the cause. The husbands of the most involved women took up some of the slack at home, and some of them were actively involved in the organization. It would be more accurate to say that the core group *lived* their occupation rather than worked at it. They were determined. They pursued narrow goals relentlessly. They often said that they were fighting for their families. In truth, they were formidable foes.

Because they did not have a great deal of money, they were forced to use all their resources to meet dozens of challenges, and they succeeded to an extent that surprised even themselves.

The Leader: Lois Gibbs

Lois Gibbs, the association president, commanded the respect and loyalty of the people who worked side by side with her for the many months that it took to achieve their goal.[9] The story of her personal growth from a shy, inhibited person—afraid to speak up in public, apolitical, not interested in community affairs—to a very capable, nationally known woman is detailed in her own book (Gibbs, 1982). Her loyal coworkers saw her as their leader, the one who had all the information at her fingertips, who knew all the people to contact in the government bureaucracies and in the press, who knew how to talk to the appropriate people, and who knew how to speak up when necessary. She was the one who could make final decisions, plan strategies, and come up with new activities and new approaches to persistent problems.

It was far more than the small core group that Gibbs inspired, however. She was able to gain the confidence of a good part of the larger Love Canal community. From the very beginning, she won the respect of political figures and the trust of the working press. Although some residents of Love Canal did not agree that the Homeowners Association was necessary, that it sought universally accepted goals, or that it sought them in a manner approved by everyone, there was general agreement that Gibbs was a powerful leader and that the Homeowners Association was the prime citizens' group. The Homeowners Association consistently listed more than 600 families as members, and Gibbs and the association were seen by hundreds of people as their chief representative to the outside world of news media, the state task force, governmental and private agencies, attorneys, helpers and would-be helpers, casual callers and curious visitors, film makers, sympathizers from various places, and even the tiny collection of "kooks" that any publicized event seems to attract.

There were days when representatives of all these different groups called or came to the association headquarters, many at the same time, so that Gibbs and those who worked closely with her would turn from comforting a distraught mother, to talking to a visiting reporter, to finding someone to guide interested college students, while asking a politician's secretary to wait on the phone.

Gibbs set the tone for the office. She greeted callers pleasantly, acted calm in the midst of the days of confusion, and had an array of jokes and pithy comments for other days—the long, quiet ones, when hopeless despair permeated the air and the question in everyone's mind was "Will we get out before it's too late?"

Making the Most of What You Have

The association's most active workers started out with everything to learn, and they learned quickly to press their cause in every way possible. As they acquired skills and knowledge, legal, educational, and political strategies all served their purposes.

One of the association's first acts was to hire a lawyer, who then advised the association as a whole and also represented several hundred families in private lawsuits against the city, the school board, and the Hooker Chemical Corporation.

The association leaders educated themselves about toxic chemicals, about pertinent local and national laws of waste disposal, and about political processes. They did health surveys, and developed some sophistication in understanding what constitutes good science. They educated themselves and spread their message to others by speeches—locally and in other parts of the country—by telephone calls, and by interviews with the press. They answered all letters (more than 3,000, from every state and several foreign countries). They kept a register of residents, maintained clipping files, wrote a chronology, knew a great deal about every family that belonged to the association. Most important, they absorbed and retained everything possible about the details and progress of their cause. They attended all public meetings remotely connected with their interests, testified before investigatory bodies, and exchanged information with other groups that had similar problems.

The association workers contacted and kept in constant touch with all elected political figures who could help them in any way at all. The mayor, county legislators, state representatives, congressmen, senators, and the president of the United States frequently heard from Gibbs and from many other members of the association.

They marshalled and husbanded their resources carefully. Although they were happy to get the endorsement of other groups, they did not dilute their cause by taking on more than the issue of the effect of toxic chemical wastes on people. In general, they used opportunities skillfully—putting anyone to work who could help in any way and using chance encounters as opportunities to state their problems and viewpoints—and they utilized all tangible materials to their fullest.

Most important, the workers gave of themselves. They provided a constant presence; they persisted; they were always there. They did all the tedious, laborious tasks familiar to anyone who has ever worked in a volunteer organization. They created and participated in events—and scarcely a week went by without events. Although many of the events—the meetings, the walks, the demonstrations—drew only a few Love Canal residents, the events were constant. There was always some activity going on, some feeling of movement, something to boost the morale of the members, something for the pens and the cameras of the news conveyors to record.

Mass-Media Involvement

The combination of news people and news products that are popularly referred to as the mass media—newspapers, popular magazines, television and radio news and other programs, and the news gatherers, reporters, and producers—played a critical role at Love Canal and have been mentioned as part of the scene and cited repeatedly throughout this book. Local, national, and international print and broadcast media were all important.

The work of the local newspaper and television reporters was especially remarkable, for they not only alerted the people of Western New York to the Love Canal problem but also provided a continual stream of information about all aspects of the problem. Because they were located in the area, they were concerned about the long-lasting issues, steeped in the details, familiar with all the involved people, and accountable both to them and to their other local readers. These journalists could take time to explore some issues in detail over a prolonged period of time. Furthermore, some local reporters played roles beyond reporting news. *Niagara Gazette* reporter David Pollak, for example, was not only responsible for some early articles about Love Canal (written alone or with David Russell) but also traced the source of chemicals in a home sump pump to the Hooker wastes. Mike Brown, another *Niagara Gazette* reporter, was concerned about the existing health problems at Love Canal, the paths of chemical migration, and the potential health affects. He spoke with many residents, arranged meetings, published anecdotal accounts of illnesses, hung onto the Love Canal story tenaciously, and within a few years had expanded his interests to toxic-waste problems nationwide (Brown 1979). It apparently was not easy for the *Niagara Gazette* reporters at first. In his book, *Laying Waste*, Brown speaks of the problems he encountered as a chemical-company-town news reporter who was obsessed with the Love Canal story: "In my own newsroom, there seemed to be an unwritten law that a reporter did not attack or otherwise fluster the Hooker executives" (Brown 1979, pp. viii, 25, 50; see also Swan 1979).

In the autumn of 1978, Michael Desmond of the *Courier-Express* wrote a series of articles about chemical-waste disposal that not only won him professional accolades but also a special award of merit from the EPA's Region II, on June 27, 1979. The award stated: "As a result of his investigation, the EPA was prompted to move up its projected date of issuance of hazardous waste regulations and to assign more manpower to handle the hazardous waste problem" (Desmond 5/11/81). Congressman LaFalce, in calling for EPA regulations, inserted the Desmond series in the Congressional Record (*Courier-Express* 9/24/78).

Local reporters cited throughout this book kept the story in the consciousness of readers on a daily basis. There was also a good deal of local television news coverage. Miranda Dunne (ABC-TV) and Marie Rice (CBS-TV) followed every event at Love Canal, and others reported frequently as well. Paul MacClennan's weekly environmental column in the *Buffalo News* focused repeatedly on Love Canal, particularly on the government's performance, with frequent commentary on the Hooker Company as well. The editorial pages of the three local papers regularly carried both editorial and popular opinions about the situation.

The local news people were joined by influxes of reporters and cameramen from the national and international wire services and the national television networks whenever there were dramatic developments. The accounts not only alerted the nation and the world to the Love Canal problem, they also alerted the Love Canal residents to the hazards they were encountering, and they heartened the organized residents, who considered the working members of the mass media sympathetic to them.

The power of the mass media was in arousing public opinion and keeping the spotlight on decision makers. These decision makers wanted to look good to the voting public as they made their decisions, based on legal mandates, resources, their personal feelings of interest in the toxic-waste problem, or their compassion for the Love Canal people.

One of Wayne Hadley's first instructions to Lois Gibbs had been, "Learn to work with the media." In fact, all the actors in this drama wanted access to the news media to publicize their views. Senior government officials, such as the governor and the health commissioner, could talk to news conveyors at any time they chose with reasonable assurance that their views would be presented. Government agencies have public relations departments ready to supply press releases in the standard formats and to cooperate in other ways with the news people. Lesser government officials and workers were sought out by the press for their opinions and for explanations of their work. After a short time, their public statements became more infrequent, more circumspect, more in line with established departmental policies. On occasion, "gag orders" were issued to government employees, such as those in Niagara County and the city of Niagara Falls (Silver 4/13/79). The EPA was also accused of this (Dearing 5/22/81).

The Hooker Company officials controlled fairly well what was printed about the company. Little appeared in the press expressing Hooker's views for several months after Love Canal was declared a health hazard. Then, on March 15, 1979, the company called a press conference, distributed written statements (Francis 3/16/79), described the conference in full-page advertisements (see Hooker 3/19/79), and thereafter made their views known through full-page newspaper advertisements, colorful brochures issued through their public relations offices in Niagara Falls and Houston, Texas, and through occasional careful statements to the press. Their advertisements and *Factline* brochures (also distributed to their own work force) stressed their care and concern for the people of Niagara Falls (see Hooker 5/29/79, 7/13/79); the fact that they are important local employers (Hooker 12/14/79); and their adherence to the state of the art of waste disposal (Hooker 3/19/79, 9/5/79).

As time went on, Hooker asserted not only their blamelessness for use of the Love Canal site after their sale of it (Hooker 9/5/79) but also the unfairness of being blamed in light of their display of responsible behavior in warning the school board of Niagara Falls about the chemicals in the ground when they sold the site to them for one dollar (Hooker 5/29/81). After February 1981, the company widely distributed a magazine article favorable to their position (Zeusse 1981) and called it to the attention of network television and radio outlets. The article's author, Hooker president Donald Baeder, and reporter Mike Brown appeared on ABC's *Nightline* (5/21/81) in a discussion of the media's role in presenting Hooker's position.

In contrast to government and corporate officials, who apparently had the attention of the press when they wanted it, the Love Canal people had to learn—and did learn—how to attract and maintain press attention when they wanted to keep their plight public, wanted to get their point of view across, or, sometimes, wanted public officials to make commitments to them in a public forum.[10]

As described in an earlier chapter, Gibbs learned early in her leadership to say things in a direct, forceful way and to ask questions and make statements shortly after public meetings started, in time for the reporters to file their stories. One of the Homeowners Association's first acts was to elect Gibbs as the official spokesperson for the association. This meant that, although all residents were free to tell their own stories and offer their opinions, only Gibbs spoke officially for the organization. (This power was challenged later, as we shall see.) Given the beliefs shared by so many, the residents' basic messages were fairly consistent, even though many people spoke to the press. Love Canal residents provided a wealth of individual human-interest stories, and many overcame their initial reticence and learned to tell their stories concisely and precisely.

For months, Gibbs's day began with reporters' telephone calls in the early morning hours, and she answered calls all day long. She tried to have something of interest to tell reporters, even if it was only that the Love Canal residents felt neglected and that, in their opinion, nothing was happening.

Gibbs and the core workers, who also were interviewed frequently, were careful to stay on good terms with the reporters, chiefly by understanding what was important to them as working people. The association leaders alerted them as much as they could to ongoing events (and in turn were alerted by reporters); and, although they kept various reporters' special interests in mind, they tried to avoid favoritism in supplying news releases. They tried to be absolutely accurate in their statements to journalists. They knew that inaccurate reports created trouble for reporters, who would then see them as unreliable informants and might even "cut them off." Gibbs and the other workers did not let factual errors or statements they thought unfair go unnoticed, however. They would call responsible reporters immediately and point out their errors or present a different interpretation. In short, both the reporters and the association leaders had something the others wanted, and they were linked in a relationship of mutual benefit and control.

During the period from 1978 to 1981, news reporters spoke and wrote millions of words about Love Canal and its people, about the activities of the government and the organized residents, and about the entire problem of toxic waste disposal. The residents' strength was derived not only from their organized efforts. In part, the reporters helped make the people powerful by amplifying their voices and keeping their cause before the public.

Differing Perceptions

Thus far, we have considered beliefs shared by the organized residents and some tactics that stemmed from those shared beliefs. All the Love Canal residents were not swiftly welded into one strong and loyal band of outraged citizens, however, so we must consider the social environment of the Love Canal Homeowners Association more carefully. The behavior of various governmental and other forces provided part of that environment, but it is important to note that not all the people in the ten-street-wide area agreed by August 4, 1978, or later, that they shared common problems. (See also Hufsmith 4/19/79.)

Given the nature of the Love Canal as a disaster, it was not surprising that there were few unanimous perceptions among the people.[11] In more familiar types of disaster there is a moment of impact—when the dam bursts, the volcano erupts, or the tornado touches down. Although controversy may arise later, there is agreement among people—whether or not they are affected—that something potentially harmful did occur. At Love Canal, there was no moment of impact when physical surroundings changed suddenly. Just as responsible agencies followed the pattern of ignoring, denying, and minimizing, and just as private sources did not rally around an almost invisible event, so the nature of the occurrence affected the reactions of the residents themselves. The chemicals had been present for some thirty years by the time the Love Canal was declared a health hazard. For community members, the mental process of moving from a condition of ignorance about the chemicals, to belief in the possibility of personal danger from them, to sharing the beliefs of others and joining in activities—some of them unaccustomed ones—with organized residents was rather slow and uneven overall. It affected a few people at first, then picked up speed as the event was publicly defined as dangerous by the government, by the mass media, and by fellow residents.

Some of the factors related to the variable rates of movement from a state of ignorance to a state of shared beliefs, have been mentioned earlier. Some people did not live in places where anything untoward seemed to be happening. Other people who did notice that something was happening, some as early as the 1950s, did not recognize that these events might indicate serious dangers. When such thoughts crossed their minds, many felt reassured because no one in authority told them that there were problems. Some were directly assured by authorities that all was well, some had been told that the fields near the school were going to be turned into parks, and others inferred, from the presence of the school and the fact that they had been granted mortgages, that the appropriate authorities had examined the area and decided that it was safe.

Because Niagara Falls is a chemical-manufacturing town, various effects related to chemicals—odors, burns, skin irritations—seemed unremarkable to the residents. Some felt inhibited in admitting even to themselves that chemicals, the source of their livelihood, could be causing them personal problems. Some felt even more inhibited in admitting suspicions to others, let alone engaging publicly in organized activities concerning the dangers of chemical products. So the presence of chemical companies as knowledgeable authorities and as forces to be reckoned with played some part in the progress of people's thinking along the route from ignorance to shared beliefs with other residents.

Love Canal residents believed differently, perceived differently, and acted differently depending on the age of the head of the family, whether or not the family lived on the inner or outer rings, and whether the family rented or owned their home. Because perspectives differed among the residents, the leaders of the Love Canal Homeowners Association grappled with problems of legitimacy—that is, of representing the views of the whole community to the outside world and to the community itself. The leadership also had to find strategies and tactics that would maintain the loyalties of the mass of the membership, that would not push away those who might be repelled by too much activism or those who would feel impatient with too little.

Most social movements encompass diverse constituencies, and often the different groups separate or organize separately and spend as much time and energy fighting each other as they do fighting the common enemy. The situation was somewhat different at Love Canal. The leaders of the several groups were aware of the dangers of factions and splinter groups, and they knew they had to avoid the worst splits. Most felt they were in a desperate struggle to protect their own health, that of their children, and their major savings, as represented by their equity in their homes. As a rule, the groups presented a united front to the world. As Gibbs reminded her group's members and the leaders of other groups:

> We're a family. We can have our fights but we stick together against our enemies. Remember, we have to work together to get out of this mess.

Age as a Factor

Although there were people of all ages who were certain that they had problems related to chemicals and wanted to get themselves and their families out, there were also differences related to the age of the family head, which is itself an index to the stage of the family's life cycle and their perceptions of their own resources and future prospects. [See also Rowe (Masters) 4/19/79.]

The oldest people (those aged sixty and over) were more likely than anyone else to say that there was nothing wrong at Love Canal. They would point out that they were living proof that there was no danger from the chemicals or, alternatively, that, if there was danger, there was little to be done, for they were a group whose resources tended to be limited. Talking about the problem and continuing to publicize it did no good in their view, caused their property values to decline, and contributed to the neighborhood's deterioration. If anyone blamed the department of health and saw the mass media and the strongest citizens' organization as troublemakers who had created the problem, that person was more likely to be found among the older group than among the others.

The youngest families, people in their twenties and thirties, were most eager to remove themselves and their children from the area. They perceived that danger applied very directly to their children and to themselves, and they also had more possibilities of starting anew than did the oldest community members.

The families in the middle, those between forty and sixty—some the parents of late adolescents—had yet another perspective. These people had completed the early years of hard work and childbearing. Many felt they had reached a point in their lives when they could anticipate some relaxation and individual freedom, with their family responsibilities and financial cares lessened. If they were homeowners, they thought of their homes as places they could enjoy now more than in the past, with mortgages well along toward completion or with payments easier to handle at this stage. Some had planned to sell or rent their homes while they traveled modestly or moved to smaller quarters. Some dreamed of moving to warmer climates as soon as they could retire. Now their plans were dashed, and they were weary at the very thought of moving. Although they wanted to be physically out—if the situation were dangerous and their homes were now worthless as investments—still, many would have been happy to stay. If they could have been completely assured that their Love Canal homes were as safe as they had previously assumed they were, then they could live in them without fear and could have the financial security they had anticipated from home ownership.

There were some people in all age groups who shared all these sentiments, but there was a rough correlation between different ages and differing viewpoints. The important point, however, is that people's perceptions were determined by the perspective from which they viewed the very puzzling phenomenon of Love Canal-as-disaster, and perceptions influenced behavior.

Some Inner-Ring and Outer-Ring Issues

One of the first questions the Love Canal Homeowners Association addressed at its first organizational meeting was who the organization represented. Based in part on the addresses of the people attending the meeting, the Love Canal neighborhood was defined as extending from 93rd Street to 103rd Streets from west to east, and from Colvin Boulevard to Buffalo Avenue from north to south.

Several weeks before Lois Gibbs started her campaign to close the 99th Street School, a small group of people whose homes directly abutted Love Canal had discussed organizing to withhold tax and mortgage payments on their properties. Some members of that early, informal group were elected officers of the Homeowners Association at its first formal meeting. When Congressman LaFalce announced that $4 million in federal funds would be devoted to the Love Canal problem, many people mistakenly thought the money would go directly to residents. One association officer who had been active in the earlier group announced the intent to form a subgroup of the larger association, to take advantage of the opportunity for immediate relief for the most affected residents (Safranek 8/6/78).

The organizational issues concerned not only legitimacy (whose interests are represented and by whom) but also basic strategy (whether to solve the obvious problems of a few quickly or form a large, continuing organization). Gibbs and Cerrillo expressed the problem as one of loyalty or conflict of interests for the involved officers. Within a few days, the problem was solved. The new organization's members met and voted to remove from office those who wanted to form the subgroup. The latter responded by resigning from the association. As inner-ring residents, they soon moved away.

It was now clear that the Homeowners Association was to represent the larger neighborhood, not just those people whose homes were closest to the canal. In fact, once the inner-ring residents moved, the focus of organization activity changed. Few of the inner-ring residents remained active in the organization after they moved,[12] and the focus of the association's attention shifted to matters that concerned the residents left behind.

Renters versus Owners

The largest group of Love Canal residents who were not Homeowners Association members were the people who lived in the LaSalle housing development (also called Griffon Manor, or "the project"). The LaSalle development was built in 1971 to accommodate families with low incomes. It consisted of 250 garden apartments, arranged in a series of courts on long, winding streets, with a large field separating the project units from the private homes on 97th Street. There were also 54 older families living in a cluster of apartments for senior citizens.

During the summer of 1978, some LaSalle residents became aware that trouble might be brewing nearby that could affect them. As they read about the state's efforts in the Love Canal area, they became concerned that the state would attend only to the needs of property owners. At an emotional meeting in late August 1978 (Perrault 8/30/78), called to discuss the construction work among other things, Commissioner Hennessey (director of the state task force) told the audience that the governor had reminded him not to forget the renters. In response, a woman from the low-income project charged that the homeowners were being taken care of and the renters forgotten.

Soon after this, a citizens' group emerged—the Concerned Love Canal Renters Association. The group's purpose was to represent the special concerns of the housing-development residents in regard to Love Canal. This organization had problems from the beginning in carrying out its chosen task.

A few weeks earlier, another organization had emerged among the project residents, the LaSalle Development Tenants Association, concerned about a variety of problems involving the physical condition of the housing project and recreation for the children. Both groups were led by individuals who were trying to rally others to work with them. The potential members were thus divided between two groups in addition to the Homeowners Association.

Unlike the Homeowners Association members, who all shared concerns about the sure loss of value of their properties as well as about the relatively uncertain matter of health effects, the LaSalle development residents' chief issue in regard to Love Canal was only in the uncertain matter of the health effects, with the related question of whether it was safe to remain where they were. Not all LaSalle residents saw these as important questions. Some had lived in the area for many years. Others had been there for a much briefer time, did not know whether they would stay, felt less exposed to chemicals, and did not share health concerns with the longer-term residents.

Many tenants depended on public assistance and thus hesitated to complain about anything the government was involved in. All LaSalle residents, by definition, had low incomes. Although the Homeowners Association stressed that people wanted to be moved out of the area, many renters did not share that goal, because their choices were limited. Although there were drawbacks to the LaSalle project, many believed it was the best public housing project in the city, with roomy apartments, located in a suburban-like atmosphere, with good schools, and without the social hazards of many inner-city neighborhoods. Many of those whose chief concern was their living conditions were mollified during the first week of October 1978, when the county health department inspected and promised to send violation notices to the city's housing authority, which promised, in turn, to take care of the problems (Delmonte 10/8/78).

In order to create a large citizens' group, Gibbs made efforts to include the project residents in the Homeowners Association, but they remained formally separate from the larger group, even though a few LaSalle project people were association members and regularly attended public meetings of that group.

The LaSalle people were generally suspicious of the Homeowners Association from the beginning. The renters feared that the Homeowners Association would not focus on their needs. During one of the early meetings of the Association, some property owners questioned the presence of the people from the LaSalle development and whether they needed to belong to the Homeowners Association. The remarks were to the effect that the renters could move easily, in contrast to the homeowners—a perception that was not shared by the poorer people.

With 65 percent of the LaSalle development residents black and almost all the property owners white, racial sensitivities were easily aroused (see Sanders 5/1/79). At an early association meeting, racial comments were muttered or thought to be muttered, and the renters' feelings were hurt; some felt unsure of their welcome, and many thought their interests were secondary to those of the property owners. When Gibbs and Cerrillo, at separate times, tried to assist the LaSalle groups and tried to interest them in joining with the association, their visits to some early organizing meetings of the LaSalle groups were seen as self-serving attempts to increase the size and power of the Homeowners Association. They were told they were not wanted at LaSalle meetings (see Stutz 9/7/78).

The clincher came for the project residents a few weeks later, when the Concerned Renters group sought a court injunction to prevent the commencement of the remedial construction work until the LaSalle development residents were satisfied that the state would pay special attention to their needs (Shribman 10/8/78). When the Homeowners Association did not join the petition, the renters had proof that their suspicions of the larger group were correct. Viewing themselves as propertyless and powerless, they claimed:

> They forgot us low income people, they think we're trash. We're people too. We've got kids and we worry. (Shribman and MacClennan 8/29/78)

At times, the differences in income and life circumstances between the residents of the federally subsidized housing project and the largely working-class homeowners created a chasm, and they viewed each other quite unrealistically. When the news headlines proclaimed that millions would be spent at Love Canal, the project residents assumed (as did many in the Niagara Falls community) that the property owners were going to become rich as a consequence of the Love Canal tragedy.

The outer-ring homeowners, however, knowing that the areas of evacuation had been set arbitrarily and fearing great financial losses, were upset at the thought that government resources were to be spent on people who did not pay property taxes and who were in many cases already receiving public assistance. The outer-ring homeowning families envied what they saw as the freedom of renters to leave the area if they wished. The renters did not find it so easy to move, or even to consider moving, for most had few resources to go anywhere.

The state task force recognized the Concerned Renters group, provided some professional assistance in organizing, invited the group's president to weekly task force meetings and provided office space and equipment. Despite the official recognition and support for their efforts, the renters association did not become a powerful force at Love Canal. The Homeowners Association leaders, increasingly wary of the state's intentions, viewed the state's recognition and support for the renters group as part of an effort by the state to split the citizens' forces.

For the first few months after August 1978, there were occasional discussions between the presidents of the Homeowners Association and the Concerned Renters group about the possibility of joining forces formally. However, when a formal relationship appeared possible, the renters' group leader asked for equal representation for her group on the Homeowners Association board of directors. The smaller group's leader wanted to be sure her organization would not be swallowed up. The larger group, however, had evolved in a certain way, was now formalized, and would have required reorganization, at least, and certainly a loss of power and decision-making ability if it accommodated the smaller group by granting it equal representation. In addition, Gibbs was having problems with internal dissenters in the homeowners group at that time, so the idea of sharing power with what she and others viewed as a disorganized, fractious renters' group was unacceptable to her as a leader.

Although the groups never formally joined forces, they existed in a relationship of mutual acknowledgment and occasional coalition when it was necessary to present a united front. As Gibbs said later, "Some groups just can't get along in one big organization. They're just too different. But they still should try to work together, not turn on each other."

The residents' united front was actually threatened more, in Gibbs's opinion, by dissenters from within the Homeowners Association. They influenced the course of that large group, while the association's leaders learned to steer a middle course, to keep the association together.

Dissent from Within: The Action Group

The outside world granted recognition to the Homeowners Association and to Lois Gibbs rather quickly. In fact, recognition was almost thrust upon

Gibbs. The trips to Albany and Washington, the attention from reporters, the private meetings with the governor and other high officials, participation in the task force meetings, and the granting of an office were all early signs that the outside world recognized and accepted the existence of the association and Gibbs as its chief.

The government officials, the private agencies, the reporters, and the general public needed some *one* to talk with, for they could not work with hundreds of people on an individual basis. The residents, too, needed one trusted source of information. From the beginning there was little, if any, serious question by outsiders that the Homeowners Association represented the majority of the owners of Love Canal homes and that the association leaders represented that group adequately.

The question raised by some association members, however, was the manner in which the major goals of the group were pursued. One small band of critics began to call themselves the Action Group after a time, and they made their views known in a forceful manner.

Looking back in the spring of 1981, Gibbs noted that, like so much else at Love Canal, the unclear nature of the event itself, coupled with the vacillating behavior of the government agencies, played important parts in the ability of the citizens' group to work harmoniously:

> No one was really sure of how to reach our goals. Everyone was trying every way they knew. During the times when not much seemed to be happening, we caught a lot of flak from the Action Group. They weren't a big group but they were big enough, and important because they were some of the active members during the first few months.[13]

Harnessing the creativity and energy of dissidents who may also be the most active members of a volunteer organization, while not allowing their dissent to break the larger group apart is a problem for all such organizations.[14]

For several reasons, Gibbs and the other Homeowners Association leaders tried to incorporate the Action Group by giving them positions and tasks in the association and accommodated the Action Group's demands for confrontation tactics. They did not want to lose the help and energy of association members who were willing to work. They were very determined to maintain a solid front of residents—or at least the appearance of one. During the brief period when he was helping her to learn the ropes, Hadley had predicted to Gibbs that most residents would be moderate in their approach, but that a few would make "way out demands in a way out way." He advised her that such people were useful in fighting the establishment, because they made the more moderate demands and tactics of the larger group seem far more reasonable.[15] Furthermore, the association leaders were always a bit apprehensive that even a small group of desperate people might make real some of the almost idle threats of violent actions—burning the houses down, planting explosives, slashing tires on the construction trucks, or other actions. The association hoped to prevent such individual acts (for which the association might be blamed) by substituting more moderate, controlled demonstrations. Most important, everyone knew that the people in the Action Group were suffering, too, and were simply more vocal in expressing the despair, the anger, and the frustration that was shared deeply by most of the residents of Love Canal. In short, the Action Group was a part of the community the Homeowners Association leaders represented.

Public Protests

At various times, the association members marched on City Hall, greeted President Carter and Governor Carey at the airport with protest signs held high, picketed the construction site, and burned both the governor and the health commissioner in effigy (*Buffalo Evening News* 4/17/79). These and many other public actions served several purposes. They were genuine outlets for angry feelings; they provided a sense of doing something more dramatic than the laborious tasks carried on in the Homeowners Association office; they offered an opportunity for participation by many of the members; and they attracted media attention, which was important to the association.

At first, the association leaders merely tolerated the behavior of the Action Group or reluctantly threw the weight of the association behind them, but with private misgivings. They feared possible adverse reactions from the Niagara Falls community, accusations of publicity-seeking through unseemly behavior, and the loss of approval of most members of the association, who did not participate in public, planned demonstrations.

As time went on, however, for several reasons, the association leaders participated more and more wholeheartedly in public demonstrations. They felt they had little to lose and everything to gain by keeping their plight public. They became increasingly convinced that the important decisions being made about their lives were in fact based on power and political considerations. Public protests could show power and attract press attention, so that they could remind the public of their specific grievances. Even after the Action Group largely withdrew from the Homeowners Association in the early autumn of 1979, the association leaders and members planned and participated in various demonstrations and public activities when they thought it would help their cause.

Gradually, the Homeowners Association headquarters was filled with signs from the demonstrations and with other slogans expressing the group's anger about their condition. To raise funds, the association sold

T-shirts with the slogan, "Love Canal—Another Product from Hooker Chemical." Wearing these, and with red carnations in their hair or in a shirt pocket, moving about in the slogan-festooned office, shopping, and carrying on everyday activities in the community, the people provided a constant picture of protest.

Attempts to Take Over the Leadership

Although Gibbs and the other association leaders tried to accommodate them, the Action Group remained disgruntled and discontented with the association leadership, for they had proof every day that they were *not* getting help they wanted from the government. The discontented group had already joined in some actions independent of the Homeowners Association leaders by the time the picketing action started on December 8, 1978. On that day, an interested, well-educated nonresident, experienced in organization work, joined the picketers. He found the disgruntled, frightened, frustrated Action Group almost immediately and became their special friend. He paid one dollar that day to become an association member. Once he was an association member and a friend of the Action Group, he helped them focus their dissatisfactions more sharply on the association leadership. During the weeks of the picketing action, rumors began to circulate, casting aspersions on the motives, abilities, and, perhaps most important, the integrity of the association's leaders, hardest workers, and consultants. A few Action Group members now refused to abide by the earlier decision that only Gibbs would speak officially with the press. They called reporters, gave them news announcements, and, when some of these turned out to be neither new nor accurate, the association lost some credibility and Gibbs had to spend time convincing the reporters that it had all been "a misunderstanding."

Under the nonresident's tutelage, the Action Group tried to undermine what they decided was an elitist ruling group in the association by preparing formal by-laws to reduce the role of the elected association officers to office managers, while elected committee chairmen would become a policymaking board. The preparation of the by-laws took a good deal of time and energy, and the discussions about them sometimes turned into acrimonious personal attacks. After one such session, involving a dozen of the most committed workers, Gibbs was afraid that people would simply refuse to continue to work in the association office because of the insults they had been subjected to. Eventually, new by-laws were accepted at a general association meeting that was packed with Action Group members, who were certain that the new by-laws would create an organization that would somehow get them moved out of the area. The new by-laws did formally establish participation by more members. It was soon obvious, however, that the level of participation was to remain what it had been all along. Like most volunteer organizations, there was a chronic problem of a shortage of labor—getting people to do the many tedious and time-consuming jobs that had to be done and were done only by the most committed people. Changing the arrangements on paper did not solve that perpetual problem.

For the Homeowners Association leaders, the period from October 1978, through September 1979 seemed to be one of continual struggle with the Action Group. Their accusations and criticisms were a constant annoyance and a constant concern to Gibbs and the other core workers.

The association leaders learned several things from this experience, however. They were forced to overcome their early reluctance to engage in planned protests. They also learned that unexpected powerful, negative feelings and forces could be unleashed in the course of such protests. They learned that, difficult as it was, they *could* live with the criticism. They learned to evaluate the criticism, separating out the style of criticism from the substance and adopting and adapting the creative suggestions. They also learned to control the opposition group through their own creative use of the rules, and they learned when to go along, when to compromise, when to assert control openly and firmly, and when to call bluffs by calling for votes of confidence. They also learned to allow full membership, with all rights and privileges, to neighborhood residents only, and to limit outsiders by allowing them only honorary memberships. In short, Gibbs and her closest working companions learned the give-and-take of working on a daily level to represent and manage a variety of interests.

The Injunction Attempt, June 1979

One of the Action Group's suggestions was to prove helpful in achieving the Homeowners Association goals, although the association leaders were certain at first that it would destroy the association. At the association meeting in May 1979, when the by-laws came to a vote and the Action Group had all its members out in full force, a motion was passed in the open meeting to file an injunction to prevent the state from completing the final phase of remedial construction work at Love Canal, on the grounds that the construction work was dangerous for the residents living in the outer rings (Shribman and MacClennan 5/29/79).

Gibbs and the other leaders were concerned that, by filing the injunction, they would alienate the moderate association members and destroy the fragile working relations that still existed with many of the state personnel. They also feared that, if the motion were rejected by the court, it would result in a loss of momentum for the association (see Shribman and MacClennan 5/29/79). On June 8, 1979, however, the injunction petition was heard in the state supreme court in Niagara Falls. Although the Court refused to issue an injunction stopping the work, it ordered that the work proceed with adherence to strict safety regulations and endorsed the state-proposed plan for evacuating residents if their health was being affected by the construction processes.

It was as a consequence of this decision that some 125 families lived in Niagara Falls motels at state expense for several weeks starting in late August, 1979 with an important consequence for the residents' organizations. Within a few days of moving to the motels, the residents not only needed physician's certification that their health problems were related to the construction work, in order to stay out of their homes at the state's expense. The other immediate problem was the acute shortage of motel and hotel rooms on the long Labor Day weekend in a city housing one of the major tourist attractions in the United States.

Coincidentally, at this time, Gibbs had gone to Washington, D.C., seeking help from federal sources. She had arranged appointments to visit elected representatives and officials in every federal agency she could think of, in order to describe personally the plight of the Love Canal residents (Powell 9/6/79).

There was therefore an acute and sudden need on the part of many Love Canal residents. The major group leader, Lois Gibbs, was unavailable; and a competent person, experienced in disaster situations and with connections to a number of resources, was needed. Sister Margeen Hoffmann took her first publicized action as the Ecumenical Task Force (ETF) executive director when she confidently stepped into the breach.

Under Sister Margeen's direction, temporary housing was found for more than 300 residents during that Labor Day weekend at a nearby parochial boarding school, and arrangements were made for emergency meals and other necessities. Her behavior, her use of resources, and her sheer presence on the scene in the hour of need, while Gibbs was in Washington working on long-range problems, was the final confirmation for the Action Group that there was better leadership to be found elsewhere.

During the early days of motel living, the Action Group formed the nucleus of a new residents' organization—People for Permanent Relocation—with both renting and homeowning members. After a few weeks of sporadic activity, that group's members largely devoted themselves to the ETF's activities. They became the ETF's loyal supporters, and for a time provided the bulk of the first active resident members for that organization. The ETF organization took vigorous action in speaking out on behalf of the residents, offering them some counseling, and acting as an additional recognized organizational representative of the Love Canal people. The Action Group members were not heard from as individuals in the Homeowners Association once they began to work cooperatively with the professional

ETF organization. (Coincidentally, a few moved out of town at about that time.)

There were two distinct views about the ETF as a representative of the Love Canal neighborhood. Some people (for example, the Homeowners Association core workers) felt that nonresidents who were merely working at a job of organizing could not truly represent the residents, because they could not truly understand the experiences and perceptions of the victims of Love Canal. The other view—held by the ETF board, the executive director, the EFT adherents, and some government officials—was that precisely because they were not involved in the legal conflict (as every resident was assumed to be), they could best remain objective and serve the community as an addition to the resident-led groups.[16]

As for the loss of the Action Group, the Homeowners Association leaders found it easier to proceed after the autumn of 1979 without the drumbeat of criticism and the consequent drain on their time and energies. The association workers continued to use planned, public events as one part of their tactics, but they chiefly continued, for yet another year and more, to do what they had been doing every day for the previous year. Someone unlocked the office doors by 9:30 every morning, put on the first pot of coffee for the day, and then everyone worked hard all day long, doing a variety of tasks that would possibly lead to their deliverance from what they increasingly saw as a sort of hell on earth. The Action Group became an unlamented memory.

The End of the Road?

By bureaucratic clocks and calendars, the Love Canal problem encompassed a fairly short time; for the people living through the experience, however, fearing for their families' health and their financial futures, the struggle was both long and wearing. The major goals of permanent relocation seemed tantalizingly close on four occasions, which I will summarize here with a description of the association leaders' reactions.

Motel Living

From late August 1979 through the first week of November, more than 125 families lived in motels. During this period, the governor and legislature seemed to have worked out a means to move out those who wished to leave the Love Canal neighborhood. (MacClennan 10/18/79). The possibility was rumored for a few weeks. Then, late on a Friday afternoon in October, State Senator Matthew Murphy telephoned Lois Gibbs with the good news that a special appropriation measure would be introduced into the Legislature. Within a few minutes, a joyous celebration was under way in the little house that since May 1979 had served as the Homeowners Association headquarters. People hugged and kissed each other, their numbers swelled by neighbors who streamed to the association office after they heard the news on the radio or spotted a news helicopter's arrival. The reporters and television cameramen arrived in time to record Debbie Cerrillo pouring champagne over Lois Gibbs's head, amid a laughing crowd of celebrants, who cheered and felt like real champions (see Dearing 10/27/79).

That celebration turned out to be premature; over the next eight months, nothing happened to implement the promise of revitalization.

The Chromosome Uproar

On May 17, 1980, the EPA announced that a study showed there was a high rate of chromosome damage among a sample of Love Canal residents. (The ensuing furor was described in the previous chapter.)

On May 21, 1980, President Carter declared that Love Canal constituted an emergency and that more than 700 families would be temporarily moved. An EPA official phoned Lois Gibbs with the good news of the press release, and within a few minutes, a quiet celebration was under way on the lawn of the Homeowners Association headquarters. People hugged and kissed each other, their numbers already increased by the reporters and cameramen, who again recorded Debbie Cerrillo pouring champagne over Lois Gibbs's head (Ackerman and Lynch 5/22/80). This time, however, the crowd was more subdued. In the ensuing weeks a flurry of backpedaling ensued, with state and federal government proclaiming that the Love Canal people should be helped, should be moved, but that some *other* level of government should pay for such a solution.

Congress Enables the President to Act

On the last night of July, 1980, only one day short of the second anniversary of Gibbs' first visit to Albany with Debbie Cerrillo and Harry Gibbs, New York Senator Jacob Javits telephoned Gibbs to tell her the good news—trumpeted in the papers the next day (MacClennan 8/1/80)—that the federal government had offered the state of New York a loan of $15 million to buy the Love Canal homes. Within a few minutes, the inner corps of workers gathered in Gibbs's apartment, hugging and kissing each other. This time, they drank their champagne, raising their paper cups in a toast to each other in a moment caught by the cameras (*Buffalo Evening News* 8/1/80).

Within a few days, the state refused the offer of a federal loan only (MacClennan 8/5/80). Over the next few weeks, a complex loan-and-grant arrangement was finally worked out between state and federal representatives,even though it looked for a time as though this federal-state solution would die at birth.

The Presidential Race

In August 1980, the Democratic party held its national convention in New York City. Members of the Love Canal Homeowners Association demonstrated outside the convention hall, carrying inflated children's boats printed with slogans proclaiming themselves Carter's Boat People (MacDonald 8/12/80; Borrelli 8/15/80). They felt that the tiring effort was worth their while and that it played a part in the final decisions. There was unceasing national publicity about Love Canal, its people, and the political aspects of the problem and the solutions throughout the summer of 1980. After Phil Donahue brought forty Love Canal residents and the mayor of Niagara Falls to Chicago and devoted a full hour's program to their concerns (June 20, 1979), the Homeowners Association received hundreds of letters, as the show was broadcast, week after week, in different parts of the country. When letter-writers asked how they could help, the association response was, "Write to congressmen and to the president." Gibbs felt that her appearance on the popular *Good Morning America* show on September 19, 1980, and her accusation that President Carter was treating Love Canal as a political football was influential in the final outcome.

More experienced political observers thought it was more important that Governor Carey had withheld his wholehearted support of the incumbent president, who was seeking a second term, until Carey received the assurances of federal support to help him solve the problems of the Love Canal people.

The final agreement was reached when two of President Carter's advisers, Stuart Eizenstat and Eugene Eidenberg, flew to an airport near Governor Carey's summer home and met with the governor (Molotsky 8/23/80). Carey was reported to say, "This is the way government is supposed to work" (Molotsky 8/23/80). "That's our system," Carey remarked later, referring to the political context of the decision (Frank 10/3/80). To the Love Canal residents, it was reminiscent of the attention their problems had received two years earlier, in the midst of the gubernatorial campaign (Allan 2/18/79). There was general agreement, in short, that electoral politics ruled the day and influenced the making of the agreement.

When the president signed the financial agreement devised by the state and federal officials, the event was suitably showcased, early in October

1980, in the Niagara Falls Convention Center, with maximum exposure for all the political figures involved at Love Canal (Herman 10/2/80). When he made his announcements about the funds for Love Canal, President Carter commended Lois Gibbs and her grassroots organization for their part in keeping the matter of Love Canal alive for such a long time. Gibbs was invited to join the dignitaries on the stage when she arose from her aisle seat as the president commended her, walked down the center aisle, and stood squarely in front of the stage, hands clasped, gazing at President Carter. The president motioned to her to come up on stage and graciously offered her his hand. Standing at the president's side during some fifteen minutes of speeches, Gibbs made the most of that opportunity. Two years earlier, she had barely begun to understand that political issues were intertwined with the health and environmental issues at Love Canal. Now she spent her precious quarter-hour whispering in the president's ear of the Love Canal residents' need for low-interest mortgages to help them purchase new homes once their Love Canal homes were purchased by the government (Coppola 10/2/80).

Her reaction to the meaning of the agreement, however, had been expressed a few weeks earlier. On August 23, 1980, the announcement was made that the complex agreement between federal and state officials had finally been arrived at (but not yet signed by the president). There was no round of hugs and kisses among the association workers and no champagne. Lois Gibbs voiced a popular sentiment: "I just hope nothing goes wrong this time" (MacClennan 8/23/80). Nothing did go "wrong." It went according to the agreement, which represented compromises between governmental officials. The Love Canal residents found themselves still bearing a burden of costs, both financial and emotional. For many, however, the very worst was now, at long last, over.

Commentary on the Citizens' Organization

The Love Canal Homeowners Association was a true grassroots organization. It arose in a working-class community in a time of crisis, when citizens joined together because they felt that their needs were not going to be met properly by their government. The organization became a strong countervailing force in opposition to corporate and governmental interests, which would have preferred to minimize the problem once they finally recognized that it existed.

In August 1980, when it was clear that one of the major goals of Love Canal residents would be achieved—that the homes would be purchased—a journalist asked Lois Gibbs whether she felt that the Love Canal story demonstrated that "the little guy" could finally win against "the big guys."

Gibbs replied: "No, that's wrong. We're not little people! We're the big people who vote them in. We have the power; they don't."

It is not clear what Gibbs was referring to at that moment. Perhaps she was referring to the part the organized residents had played in making the issue broad in scope and public in nature, rather than narrow and quiet. Perhaps she referred to the fact that the people who called themselves ordinary citizens used the means our governmental system provides to join together, to apply pressure on elected officials, and to publicize important problems by means of a free press.

As we have seen, the outcome of the Love Canal story would have been very different for the people, and for our society, had there been no citizens' associations at Love Canal. The process of grassroots organizations making their will known is inherent in American democracy (DeTocqueville 1956). What began at Love Canal with one young woman knocking on doors grew into a force to be reckoned with by state and federal officials at the highest levels.

Notes

1. See Bucher (1957) for a discussion of blaming in manmade disasters.
2. The 239 inner-ring families were offered the following:

1. The fair market value of their homes (that is, the value of a comparable home not located in a hazardous area).
2. "Loss of favorable mortgage financing payments" (that is, a sum of money to make up the difference between the interest rates they had paid originally, and the interest rates they would now have to pay to finance mortgage loans).
3. Supplemental payments (that is, sums of money—as high as $10,000—to make up the difference between the fair market value and the cost of a comparable new house in an inflationary period).
4. Closing costs.
5. Moving costs, once or twice, as needed—from the Love Canal home to temporary housing and from temporary housing to permanent housing.
6. Temporary housing, if necesary, for up to six months.
7. Costs for utility and telephone installations in new quarters.

Items 2 and 3 were called "the supplement" and ranged as high as several thousand dollars. Two families remained at Love Canal and 237 moved out, taking advantage of some or all of the package of offers. The precise details of the benefits, in addition to the home purchases, became important two years later, in the summer and autumn of 1980, when supplemental payments could not be provided for Love Canal residents who were included in the moves supervised by the revitalization agency (to be discussed later).

3. For an early statement on the concept of relative deprivation, see Merton and Kitt (1950, pp. 42-53). For the importance of this sense of deprivation in the creation of social movements and their organizations, see Freeman (1975, chapter 1).

4. See Barton (1970), Dynes (1978), and Quarentelli and Dynes (1977) for some further discussions of community responses in disasters.

5. See Erikson (1976, pp. 186 ff.) for a more descriptive discussion of loss of neighborhood and of neighbors. The loss contributed, in his view, to the "collective trauma" caused by the Buffalo Creek flood.

6. See Tsubaki and Kayama (1977) for similar reactions in Japan, concerning mercury poisoning.

7. See Dohrenwend (1978) for the concept that the strength of the impact of stress on individuals and persistence will depend on the resources brought to bear in the "coping process." See Caplan (1976) for family functions, Fritz (1961) for the concept of "primary group solidarity among survivors" of natural disasters, and Barker (1979) for a description of stress and coping at Love Canal.

8. See Antze (1976) for a discussion of the importance of an organization's ideology in helping members to cope with their major feelings and their situation.

9. For months, the residents' emotional needs were not so much those created by a condition of acute crisis but those created by a long period of attrition, when individual group morale had to be maintained to keep the group together and working (see Lang and Lang 1964).

10. See Molotch and Lester (9/75) regarding unequal access to event-making; and see Schoenfeld, Meier, and Griffin (1979) for a discussion and review of literature on the daily press's role in covering environmental issues. (See also Ploughman 4/19/79.)

11. Materials about Love Canal residents' perceptions were obtained from lengthy individual interviews conducted in the autumn of 1978 with fifty-nine families. The interviews were conducted by the author, five graduate students in sociology, and one undergraduate. The interviews were analyzed qualitatively during a fifteen-week seminar following data collection. In addition, there were innumerable confirmations from my observations of and informal conversations with the residents over a three-year period.

12. Debbie Cerrillo, the association's vice-president, and Joann Hale were exceptions in that they remained interested and active even though they were among the first people moved out of the area permanently, in 1978.

13. Lois Gibbs, 1981. Used with permission.

14. See Olsen (1968, chapters 9, 10) for a general discussion of this organizational problem.

15. See "Mau-mauing the Flak Catchers" (Wolfe 1970) for a description of this tactic, as used by black antipoverty groups in the 1960s.

16. At the very end, when the Love Canal ordeal was over, or almost so, the Homeowners Association reduced its activities markedly, and the headquarters was moved to a member's home. The Ecumenical Task Force organization, however, with its professional director and paid staff, changed its name to the Ecumenical Task Force of the Niagara Frontier and broadened its scope to include toxic waste problems for the entire Northwestern New York area (see Hoffmann 1980). As a professional group, it had found a cause, a base, and an important need it could fulfill.

See McCarthy and Zald (1977, p. 1215) for related theoretical discussions of social movements stemming from shared grievances and those stemming from the work of issue entrepreneurs.

8 The Ending and After the Ending

The Love Canal story continues. In August 1978, New York Health Commissioner Robert Whalen recommended that "families with pregnant women" and the "approximately 20 families" living on the streets bordering the southern end of Love Canal "with children under 2 years of age, temporarily move from the site as soon as possible" (Whalen 8/2/78). In contrast to this rather modest set of recommendations, by the summer of 1981, more than 500 families had moved out—with their homes purchased by the state; hundreds more purchase applications were pending; and the large public-housing project was more than half empty. The neighborhood, once full of sound and life, was empty and silent, with grass growing tall among the boarded-up homes and apartment buildings and over the huge, gently sloping clay cover that makes the old Love Canal look like a great grave, with the abandoned school as the tombstone. Although the organized residents who preferred to move out had "won," for them and for those who remained behind there was and is no feeling of triumph—nor was the way easy, nor is the end in sight.

After the U.S. Congress approved an emergency appropriation in the summer of 1980, allowing the president to spend up to $20 million to relocate Love Canal families (U.S. Congress 7/3/80), there were weeks of negotiations between the Federal Emergency Management Agency (FEMA) and the state task force representatives. They finally agreed that the federal government would lend the state of New York $7.5 million at 8.25 percent interest and would grant them another $7.5 million, while the state provided $5 million in revitalization funds—earmarked by the New York legislature and governor nine months earlier for Love Canal (*Buffalo Courier-Express* 7/3/80).

The final arrangement was complex, and it was not lost on the residents that the negotiations took place against background of conflict between the governor of New York and the incumbent president about the amount of support the governor would give to the president in the upcoming 1980 election. One news headline summed up what lay in store for the Love Canal people: "Congress, White House at Odds over Intent of Love Canal Funding" (Peck 7/3/80). The agreement was finally signed on October 2, 1980. Home purchases and neighborhood revitalization were then administered through the Love Canal Revitalization Agency, an entity established June 4, 1980 (N.Y. Laws 1980; see Westmoore 7/1/80).[1]

It is beyond the scope of this chapter to describe all the difficulties that agency was to face. They worked to satisfy conflicting mandates. The $20 million was to fund not only home purchases for hundreds of families and relocation expenses for the renters, but also revitalization of an area ten blocks wide. They performed these tasks in the limelight, working with residents who were angry, worried, confused, and "totally sick of the whole thing," and who did not trust the revitalization agency. The agency's intitial interpretation of the rules for assisting residents did not help to develop trust.

By the terms of the complex federal-state loan-grant agreement, only $2.5 million of the $20 million in combined funds were available for all purposes other than purchasing homes whose owners were living in them. The $2.5 million had to be stretched to cover, among other items, temporary housing costs, the needs of the renters, neighborhood revitalization, and payments for commercial properties. Owners of commercial properties—such as the woman who had her life savings in and earned her living by operating a local restaurant and a few property owners who not only lived in Love Canal homes, but had invested in others as rental-income properties—could not understand why their needs were any less and why their properties were not simply to be bought as the homes were. In addition, approximately thirty families experienced intense anxiety and confusion when they learned their homes were listed as commercial properties. One woman, for example, had fled the neighborhood with her sons in the summer of 1978 in order to protect the children. She had rented her Love Canal home for $100 a month less than her costs, taking an extra job to earn the money to make up the difference, while paying rent for new quarters as well. A second young couple had not had enough money for the down payment on their home several years earlier, when they first moved in. After the wife's father purchased the home in his name, making the down payment of a few thousand dollars, the young couple made all the mortgage and other payments. The young couple—parents of a congenitally brain-damaged child—desperately wanted to move. I watched both these families as they cried publicly in their helpless anger at arbitrary rules set down by authorities working far from the realities of Love Canal. After they suffered weeks of anxiety, both families were finally told that their homes would be purchased.

Each family that moved had different problems, but some problems were shared by all, as hundreds sought homes in a relatively small real estate market. Potential buyers, accustomed to mortgage rates at 6 percent (and some dipping as low as 4.5 percent), had to cope with mortgage rates climbing well over 13 percent by the time they were ready to make their down payments. Renters had to locate suitable quarters, with limited funds. By summer 1981, many people were still hunting for new places to live. In addition to the usual problems of locating a home in a suitable neighborhood, near good schools, and within a reasonable distance from their workplaces, many residents had some special requirements. Many wanted to locate as close as possible to former neighbors, friends, or relatives. The worst problem, however, was that they were haunted by the fear that they might again, unknowingly, find themselves near a chemical-waste dump site. For that reason, many wanted to buy homes outside the city limits if possible.

The experiences of some active Homeowners Association workers are probably representative. One couple moved thousands of miles away as soon as their home was purchased. The husband had transferable skills, and they wanted "to get out of New York State forever." One couple, by a happy chance, purchased a house that the wife had known and liked since her teens, when she had worked in it as a babysitter. Three other families lived in cramped, temporary quarters for months and spent hundreds of hours trying to find homes they could afford—near the husbands' work and where they felt safe. Two of the three families finally found suitable houses similar in size to their Love Canal homes but with monthly mortgage payments that were double their previous ones. One family is still seeking a home they can afford. In order to stay within their means, one family moved to a mobile home, while another purchased a home with an elderly family member.

A minority—roughly estimated to be from 9 percent to 27 percent of the families eligible for relocation—seemed willing to remain in the neighborhood if they could be assured their homes and the neighborhood were really safe.[2] The revitalization agency hoped to resell some 300 of the homes purchased by the state (located in the outer rings and not known to be near underground swales), as part of their plan to stabilize and revive the neighborhood (MacClennan 6/20/81).

In the summer of 1980, the EPA had launched studies to ascertain whether the neighborhood air, water, and soil were polluted with chemicals. That agency had retracted an early statement that their studies would show which homes were safe. However, those hoping to make decisions based on solid information were awaiting the results of the studies, which were long overdue by the summer of 1981 (Tyson 6/18/80). The reason for the delay may have been political. The new Republican administration had not yet stated its position on Love Canal, but it had made it clear that environmental concerns were not of high priority (*Niagara Gazette* 4/22/81). By mid-June 1981, it was reported that Congressman LaFalce had written to the new EPA director, Anne Gorsuch, accusing the agency of delaying the release of the study results and stating that his inquiries, including personal telephone calls, went unanswered (LaFalce 6/19/81; MacClennan 6/20/81). One month later, "Delay Seen in Release of Canal Tests" headlined an article stating that September 1981 was the new target date for the report (Tyson 7/8/81).

The delay may have been an indication of the complexity of the task or other agency-related problems, but the burden of the delay fell on some Love Canal residents. They were intensely ambivalent, trying to decide whether they should move, while their homes could still be purchased within the terms and deadlines of the federal-state agreement, or take the chance that the neighborhood revitalization would indeed take place someday (see MacDonald 4/18/81). Rather than reassurances about the environment, however, a state health department report confirmed that dioxin was in the neighborhood creeks (NYDOH 4/81, pp. 8, 11; Allan 6/6/81), and the newspapers reported that chemical-smelling liquids were entering the basements of two homes, each several blocks from the canal site—one north, one east (Tyson 6/4/81). Rather than receiving reassurances that their interests would be protected, the people read news accounts about the revitalization agency's tentative plan, in the face of the uncertainty, to ask future home buyers in the area to sign a disclaimer of liability, so that the city could avoid further lawsuits (MacClennan 6/20/81).

When the executive director of the revitalization agency announced his intention to move himself and his family into one of the empty Love Canal homes (*Niagara Gazette* 5/18/81), it did not set an inspiring example but rather aroused a storm of criticism from residents, the Ecumenical Task Force, and the local newspapers. (Tyson 5/19/81; Dearing 5/22/81; *Niagara Gazette* 5/27/81). For Love Canal people, the announcement recalled the unsuccessful attempt by the state two years earlier, through the Urban Development Corporation, to sell some of the inner-ring homes at auction to bidders who were to remove the houses from the area (Shribman and MacClennan 5/26/79; Porter 5/20/79; *Buffalo Courier-Express* 7/14/79). Actions such as these, in the absence of solid assurances that the homes and the neighborhood were safe, seemed to Love Canal residents to defy reason and affirmed for many that economic concerns remained more important to officials than the residents' health and safety.

The effects of leaching chemicals on the health of Love Canal residents were still unknown by the summer of 1981. It was not until June 1981 that a special report to the legislature and the governor, prepared by the New York Department of Health, became available to the public (NYDOH 4/81). The sections of the report dealing with the results of the extensive health studies provided no data beyond those available a year earlier, in the spring of 1980, when the "adverse pregnancy outcomes" study was publicized (Page and Shribman 6/26/80). Those data showed excessive miscarriage rates among various groups of Love Canal residents. Although the latest report includes statements that there is "no excessive incidence" of several diseases, there are no data provided, no specific studies are cited and by late June, the studies were still in preparation (NYDOH 4/81, pp. 21-22; Vianna, 6/25/81). There was still no definitive answer to the question of whether chemicals had migrated to soil beyond the first ring of homes, but the health department was continuing to study the problem (NYDOH 4/81, pp. 12-14).

A report on cancer incidence at Love Canal from 1955 to 1977 appeared in June in *Science* (Janerich et al. 1981) and was covered in the newspapers (Allen 6/12/81; *Buffalo Courier-Express* 6/12/81). It was inconclusive for a number of technical reasons described in the report itself. The study reported an excessive number of respiratory system (lung) cancers when the census tract containing Love Canal was compared with the incidence of cancers in New York state, excluding New York City, or in other Niagara Falls-area census tracts but concluded there was no excess number of cancers overall. Despite the excess of respiratory system cancers in the Love Canal census tract, the authors state that there was no concentration of cases near the canal itself. The research team felt they could not infer that the excess was related to the presence of the chemical dump site. Given their three-year history with the health department—however, the residents took it with the proverbial grain of salt (*New York Times* 6/14/81).

In the summer of 1980, in the wake of the chromosome furor described in chapter 6, the U.S. Center for Disease Control (CDC) was promised $3.8 million for health studies at Love Canal, to be performed by the medical faculty of the State University of New York at Buffalo. The funding was available, then unavailable; the effort was on, then off, then on and then off again (see MacClennan 3/14/81). Political machinations, bureaucratic infighting, the inexperience of the medical faculty with the Love Canal problem, and the history of poor relationships between government scientists and the Love Canal community all had marked this effort. By the summer of 1981, the possibility of a health study was once again under consideration (Dearing 6/16/81; Peck 6/16/81). It remains uncertain whether the studies will be done (Peck 7/9/81), and there are no plans, in any event to do more than than a reduced version—rather than the long-term broad-scale followup the difficult situation might require.

Other uncertainties lay ahead for the Love Canal residents, who were all trying to "get back to normal living." Hundreds of lawsuits were already filed, for example, or about to be filed. Residents occasionally thought anxiously about having to testify in court and about what the outcome of their suits for property damages and injuries to their health would be and what the legal procedures would entail for them.

There were uncertainties about what would happen to all the people in the future as a result of their experiences during the years they spent embroiled in the acute public phase of the Love Canal disaster. By the summer of 1981, among the half-dozen core workers, some marital difficulties had occurred. One couple was divorced—the new roles they had played providing the final strain in their marriage. One couple, married for a number of years, was surprised at what they had learned about each other's new capabilities. They are still adjusting to their new roles and new relationship. One couple is "more in love and closer than ever." One couple, plagued by new financial worries, is suddenly arguing a great deal in a way unknown to them before. Two other couples simply feel relieved finally to have worked their way through a very difficult time: "We're just trying to lead something like a normal life again."

No one knows, of course, what their future reactions will be or what lasting effects there might be for the hundreds of other Love Canal adults—or for the children who moved from house to house, lived in motels, changed schools, and in many cases worked side by side with their parents as they struggled through years of turmoil. We only know that the story is not yet over for the Love Canal people.

The story is not over either for the rest of our society. As we look to the future, there is little reason to believe that Love Canal will be a unique event. Indeed, Love Canal has become the symbol for what happens when hazardous industrial products are not confined to the workplace but "hit people where they live" in unestimable amounts, with the potential for producing a wide variety of diseases and disorders but in such a way that it is possible to attribute chronic, slowly developing signs and symptoms to other causes.

We do not know exactly how many buried waste sites there are in the United States. Shortly after Love Canal was identified as a serious health hazard, a second task force established by Governor Carey reported the presence of 215 waste-disposal sites in the Erie-Niagara County area. Of the total—based on figures supplied by local industries—36 definitely contained hazardous-waste products and another 116 "may have received significant quantities of hazardous wastes" (Millock 1979). The EPA has estimated that there are as many as 30,000 hazardous-waste sites across the country (USHR 9/79, p. 1). The surgeon general of the United States reported to a congressional subcommittee that "it is clear that [the toxic chemical risk] . . . is a major and growing public health problem" (USDHHS 1980). The EPA also reported in 1980 that only 10 percent of all hazardous wastes were disposed of in environmentally sound ways (USEPA 1980, p. 15).

The list could go on and on. Although chemical-industry representatives do not agree with those numbers and argue for the efficacy of their disposal practices, there is little question that chemical pollution is widespread. Although scientists employed by industry and by government and even regulatory agencies may argue about the precise cause-and-effect links between chemicals and human health disorders, no one has even suggested that it is preferable to live next to a chemical-waste-disposal site.

The physical phenomena are out there, present in the world. Equally important, so are the social processes whose consequences are very real. At Love Canal, many of the most important events could be understood as the consequence of well-known social processes. In the slowly developing, multifaceted situation, the various groups viewed and defined the situation from their own perspectives, adding to the persistent conflict. Moreover, each group formed opinions and acted in terms of its self-interest—whether financial, political, professional, moral, or survival. Other social processes were going on as well. Nothing occurred at Love Canal, however, that was theoretically surprising or even theoretically new to a social scientist. What is important, rather is the fact that what happened was so broadly predictable, for it means that the same processes will be present at other Love Canals.

We must take seriously the idea of countervailing forces and self-interest. When matters of pollution arise, most people are likely to be in the position of Love Canal residents—that is, the bearers of burdens, the takers of risks that are decided upon by others, who enjoy the benefits. The Love Canal residents were concerned about survival, health, and welfare. So must we all be. The Love Canal residents learned that they had to press forcefully for even partial satisfaction of their interests. So must we all. The technical and legal means exist, or can be developed, to provide the protection we need from chemical pollutants. They must be used to fulfill the intent of preservation and enhancement of human life.

At Love Canal, although there were individual exceptions, neither government nor industry as a whole put the residents' chief concerns first. As we consider the history of Love Canal, had human values been of top priority, no company would have filled a recreational waterway with chemical wastes; no school board would have purchased nor would any company have sold such a disposal site for use as a school; no health department or municipal official would have ignored complaints; no health commissioner would have told women that double and triple risks of miscarriage were their problems; no governor and no president would have converted the removal of innocent men, women, and children from living conditions where even the possibility of serious physical harm existed into chips in a political bargaining game.

One evening in August 1978, in the midst of an impassioned protest about what many residents feared was an inadequate safety plan, a Love Canal resident cried out: "when you can hear that whistle blow, it's too late!" If we think of Love Canal as the whistle warning the rest of society about what lies ahead, in one important sense it is already too late, for the physical hazards already exist. However, if we use the warning of Love Canal to prevent or alleviate some of the social consequences that created unnecessary suffering at Love Canal, then perhaps we will have learned something from history and sociology when we face the emergence of new Love Canals all over this land.

Notes

1. The revitalization agency was headed by Niagara Falls Mayor O'Laughlin. It included Wheatfield Town Supervisor Edward Greinert, Wheatfield Councilmen James D. Heuer and Karl T. Giffin, State Urban Renewal Agency director Angelo Massaro, and Niagara Falls Councilman Joseph J. Smith. As mandated by the legislature, they chose three community representatives. They decided to select only among those not involved in litigation, which eliminated almost all the homeowning Love Canal residents. The community representatives were two ministers—one from the Ecumenical Task Force, John A. Lynch, and Rev. Dr. Leotis Belk to represent the Concerned Renters group—and Bill Waggoner, who represented Concerned Area Residents, a group of people who wanted to remain in their homes in the vicinity.

2. Interview with Richard Morris, executive director of the Revitalization Agency (6/21/81). Fifty of the 550 eligible families had told the revitalization agency by June 1981 that they were not interested in working with the agency. Another 100 families had had contracts with the agency for several months but had not notified the agency of further plans to move. I computed the percentages based on Mr. Morris's rough estimate.

List of Abbreviations

The following abbreviations are used in the text, in the bibliography, or in both.

BCE: *Buffalo Courier-Express*
BEN: *Buffalo Evenings News* (the weekend editions are called *The Buffalo News*, but all are referred to herein as *Buffalo Evenings News*.)
CDC: U.S. Center for Disease Control
DEC: New York Department of Environmental Conservation
DOH, NYDOH: New York Department of Health
DOT, NYDOT: New York Department of Transportation
EPA, USEPA: U.S. Environmental Protection Agency
ETF: Ecumenical Task Force
FDAA: Federal Disaster Assistance Administration
FEMA: Federal Emergency Management Agency
HEW, USDHEW, DHEW: U.S. Department of Health, Education, and Welfare
HHS, USHHS: U.S. Department of Health and Human Services
LCHA: Love Canal Homeowners Association
NG: *Niagara Gazette*
NIEHS: National Institute of Environmental Health Sciences
NYDOH, DOH: New York Department of Health
NYDOT, DOT: New York Department of Transportation
NYT: *New York Times*
OSHA: Federal Occupational, Safety, and Health Administration
SUNY: State University of New York
USDHEW, DHEW, HEW: U.S. Department of Health, Education and Welfare
USEPA, EPA: U.S. Environmental Protection Agency
USHHS, HHS: U.S. Department of Health and Human Services
USHR: U.S. House of Representatives

Bibliographic Note

The materials for this book derived from many sources. With the exception of confidential interviews, all materials are available to the reader. The books, articles, published government documents, and newspapers are available through the cited sources.

The newspaper articles cited are included in a collection of several thousand clippings from the *Buffalo Evenings News*, *Buffalo Courier-Express*, *Niagara Gazette*, and the *New York Times*, covering chiefly the period from August 1978 through August 1981. These have been deposited with the university archivist at the State University of New York at Buffalo. I have also deposited additional materials with the SUNY/Buffalo archivist including the following:

Government agency reports, memoranda, press releases, and other documents

Hooker Chemical Company *Factlines* and other public-relations materials

Letters written by government officials or people acting as public officers

Letters written by Lois Gibbs, Beverly Paigen, Adeline Levine, and others

Love Canal Homeowners Association chronology and newsletters

Ecumenical Task Force Annual Report and newsletters

United Way of Niagara account of expenditures of $200,000 grant

Beverly Paigen's *Notebook*, prepared for the Roswell Park Memorial Institute Board of Visitors (3/7/80)

New York Department of Health Collected Documents, 1979 (a collection pertaining to Love Canal, distributed on a limited basis in the spring of 1979).

New York Department of Health, *Love Canal; Public Health Time Bomb: A Special Report to the Governor and Legislature* (September 1978) and *Love Canal: A Special Report to the Governor and Legislature* (April 1981).

Master's projects and other materials on Love Canal that were prepared by university students and professors are available as cited or may be in the SUNY/Buffalo archivist's collection.

Other sources for materials used in the book include the Niagara County Records Department (Lockport, N.Y.); the Niagara Falls Board of Education Love Canal records; the Niagara Falls Engineer's Department; the U.S. District Court for the Western District of New York (Buffalo); and the New York Supreme Court, 4th Dept. (Niagara Falls).

References

"ABC Nightline"
5/21/81 Transcript. New York:ABC-TV.

Ackerman, Marsha
2/16/79 "Love Canal group goes to Albany; presses plea to relocate families." BCE:3.

Ackerman, Marsha, and Lynch, David
5/22/80 "Carey urges lasting relocation of Canalers: Temporary federal solution hit." BCE:1.

Adams, Stephen
12/20/78 "Meeting of December 14, 1978." Memorandum to Dr. C. Stephen Kim. NYDOH 1979, Love Canal [Collected Documents]: Sect. 47.

Albert, Roy
6/12/80 "Panel review of Biogenics Corporation study of chromosome abnormalities in Love Canal residents." Washington, D.C.: USEPA.

Allan, Jerry
8/15/78 "Canal takes on appearance of campaign issue." BEN:35.
2/18/79 "State turns cool on Canal action after campaign." BEN:A-10.
5/24/80 "Carey pushes Carter on Canal plan. Says move should be permanent." BEN:1.
10/11/80 "Panel calls Canal tests inconclusive." BEN:1.
2/8/81 "Foschio to join WNY elite on state payroll." BEN:A-5.
6/6/81 "State says dioxin is still in Canal." BEN:A-12.
6/12/81 "Canal residents called normal in rate of cancer. BEN:37.

America's Chemical Industry
11/3/80 Advertisement. *Washington Post.*

Antze, Paul
1976 "The role of ideologies in peer psychotherapy organizations." *Journal of Applied Behavioral Science* 12 (July-September): 310-323.

Averill, James R.
1979 "A selective review of cognitive and behavioral factors involved in the regulation of stress." In Richard A. Depue (ed.), *The Psychobiology of the Depressive Disorders,* pp. 365-387. New York: Academic Press.

Axelrod, David
5/12/78 "Health implications of materials found in or associated with Love Canal." NYDOH 1979, Love Canal [Collected Documents]: Sect. 10 (Initialed "DA").
7/30/79 "Interim report on N.Y.S. Department of Health environmental and epidemiological studies at Love Canal." Albany: NYDOH.
8/15/79 Letter to Gerald P. Murphy, Sect. 4 in Beverly Paigen, *Notebook,* 3/7/80.
8/21/79*a* Letter to Commissioner William C. Hennessy.
8/21/79*b* Letter to Lois Gibbs.
10/16/79 Letter to Beverly Paigen.
11/9/79 Letter to Philip Handler.

Baeder, Donald L.
12/80 "Love Canal: What really happened." *Chemtech*, pp. 740-743.

Baker, Timothy J.
8/6/78 "Experts find carcinogen at Love site. BCE:A-1.

Barbanel, Josh
5/21/80 "Peaceful vigil resumed at Love Canal." NYT:B:1.
5/28/80 "Love Canal families balk at further tests." NYT:B-3.

Barker, David
1979 "Stress and coping at Love Canal." Master's project, Department of Sociology, SUNY Buffalo.

Barron, Stephen A.
1980 "Report of pilot project: Nerve conduction determinations at Love Canal, Niagara Falls, N.Y." Department of Neurology, School of Medicine, SUNY Buffalo.

Barton, Allen H.
1970 *Communities in Disaster*. Garden City, N.J.: Anchor Doubleday.

Bartosiewicz, Thomas
3/80 "609 . . . and counting: Hazardous wastes and the public's health in New York State." Albany: N.Y. State Senate.
4/27/80 Letter to Honorable Hugh L. Carey.
5/28/80 Letter to "Dear Colleague" and Senate Resolution.

Batt, Paul
4/25/79 "Presbytery joins Canal Task Force." BEN:19.

Bazelon, David L.
7/20/79 "Risk and responsibility." *Science* 205:277-280.

Bender, Michael; Bloom, Arthur; and Wolf, Sheldon
5/21/80 "Summary report of HHS Review Panel." Submitted with a cover letter by Dr. David Rall to the USEPA.

Billington, Mike
5/28/80 "Health tests won't affect relocation Canalers told." BCE:3.
Biogenics Corporation
5/14/80 "Pilot cytogenetic study of the residents of Love Canal, New York." Prepared for USEPA by Biogenics Corporation, Houston, Texas. Washington, D.C.: U.S.E.P.A.
Blalock, Hubert M.
1960 *Social Statistics.* New York: McGraw-Hill.
Blum, Barbara
5/16/80 Letter to Barbara Quimby.
Boniello, Ralph A.
5/5/53 Letter to Board of Education, Niagara Falls, New York.
Borrelli, George
8/15/80 "Canal demonstration got point across, Gibbs says." BEN:12.
Brady, Eric
6/4/80 "Lawmaker scoffs at Roswell ruling on Dr. Paigen charges." BCE:3.
Bross, Irwin, D.J.
"Muddying the waters at Niagara." *New Scientist* 88:728.
Brown, Mike
8/10/77 "$400,000 project is asked to seal off Love Canal." NG:1-B.
2/5/78 "Red tape stalls dump solution." NG:3-B.
5/15/78 "Vapors from Love Canal pose serious threat." NG:1-B.
5/20/78 "Love Canal residents' evacuation mulled." NG:1-A.
5/21/78 "Love Canal's Homeowners plan 'loss of value' lawsuit." NG:1-B.
5/25/78 "Toxic exposure at Love Canal called chronic." NG:1-B.
6/20/78*a* "Tests outlined for Love Canal area residents." NG:1-B.
6/20/78*b* "Schools dropping role in dump remedy." NG:1-A.
6/23/78 "Army role eyed; dump fund OK'd." NG:1-A.
8/5/78 "Cheers greet word of relief." NG:1.
9/27/78 "UB Prof says Love Canal drain tile needs reviewing." NG:1-B.
10/4/78 "State, private studies at odds on Canal drain illness pockets." NG:1-B.
10/22/78 "Harvard graduate was Canal drama catalyst." NG:1-B.
11/7/78 "Regional DEC head replaced in state shuffle." NG:1-B.
11/8/78 "Dioxin feared at Canal." NG:1.
11/9/78 "Dioxin possibility deemed 'concern' but not a threat." NG:1-B.

11/10/78 "Tons of dioxin agent in Canal." NG:1.
11/14/78 "Researchers still sifting Canal health findings." NG:1-B.
12/9/78 "Dioxin found in Love Canal trenching." NG:1.
12/10/78 "State DEC refused action on Hooker." NG:1.
12/11/78 "Canal protestors arrested." NG:1.
1979 *Laying Waste: The Poisoning of America by Toxic Chemicals*. New York: Pantheon.
1/23/79 "State asked to move more from Love Canal." NG:1.

Brydges, Jerauld
8/10/78 "State evacuation plan relieves Canal residents." NG:1-B.
8/15/78 "Backyard work set at Canal." NG:1-B.
8/24/78 "Carey mixing duties, pleasure." NG:1-B.
4/27/79 "Carey: 'I can't make life risk-free.' " NG:1.

Bucher, Rue
1957 "Blame and hostility in disaster." *American Journal of Sociology* 62:467-475.

Buffalo Courier Express (BCE)
8/4/78 "No 'legal responsibility' for Love Canal: Hooker." BCE:1.
8/12/78 "Red Cross halts Canal survey." BCE:2.
8/19/78 "Ketter names environment Task Force." BCE:4.
4/20/78 "Love Canal area funding source proves uncertain." BCE:1-B.
9/16/78 "C-E reporter travels U.S. for story." BCE:1.
9/24/78 "LaFalce bids EPA act now." BCE:15-B.
11/10/78 "Toxic presence of dioxin still feared." BCE:2.
7/14/79 "Site problem for Canal homes." BCE:20 (editorial).
1/7/80 "Roswell meeting set; Dr. Paigen uninvited." BCE:21.
7/3/80 "$20 million aid bill for Love Canal passed." BCE:1.
8/12/80 "Dr. Paigen gets grant for project." BCE:1.
6/12/81 "Higher lung cancer rate found among ex-Love Canal residents." BCE:3.

Buffalo Evening News (BEN)
8/4/78 "Angry, frightened Love Canal area residents want out of danger. Scoff at Carey plans." BEN:3.
8/21/78 "UB President creates unit on Love Canal." BEN:11.
4/17/79 "Burning Mad." BEN:62. (photo).
5/11/79 "Canal work rejected for handicapped." BEN:19.
6/21/80 "US accused of hiding Canal data: Studies seen threat to suit, LaFalce says." BEN:1.
7/25/80 "Paigen gets apology of audit on taxes; Canal link is denied." BEN:1.
8/1/80 "Federal loan is personal victory for Lois Gibbs." BEN:1.

4/1/81 "Canal study office is closed officially." BEN:60.

4/2/81 "Crotty promoted on Carey staff." BEN:1-5.

Campbell, Donald T.

1979 "Assessing the impact of planned social change." *Evaluation and Program Planning* 2(1):67-90.

Caplan, Gerald

1976 "The family as a support system." In Gerald Caplan and Marie Killilea (eds.), *Support Systems and Mutual Help.* New York: Grune and Stratton.

Carey, Hugh L.

6/4/80 "Establishing a panel to review scientific studies and the development of public policy on problems resulting from hazardous wastes." Executive Order No. 102. Executive Chamber, State of New York, Albany.

Carroll, Paul

8/21/79 "Dioxin at Canal fails to reverse relocation denial." BEN:1.

Cerrillo, Debbie, and Gibbs, Lois

10/10/78 Interview with Adeline Levine.

Chapman, Verne, et al.

9/1/79 "Administrative actions on Dr. Beverly Paigen's subcontract application to the Department of Environmental Conservation." Memorandum from Council, Association of Scientists, Roswell Park Memorial Institute, to Dr. David Axelrod. Sect. 5 in Beverly Paigen, *Notebook*, 3/7/80.

Ciotta, Rose

3/15/81 "Some dream of rebuilding area." BEN:C-3.

Clement Associates

8/20/80 "The remedial construction project at Love Canal, Niagara Falls, New York: Findings and comments." Washington, D.C.: Clement Associates, Inc.

Clifford, Francis J.

5/10/78 Letter to Robert Whalen. NYDOH 1979, Love Canal [Collected Documents]: Sect. 6.

Columbia Journalism Review

1-2/81 "CBS program on Love Canal judged fair." P. 86.

Conestoga-Rovers and Associates

8/78 "Project statement: Love Canal Remedial Action Project. Prepared for City of Niagara Falls, County of Niagara, New York State, August, 1978." In NYDOH 1979, Love Canal [Collected Documents]: Sect. 47.

Cooke, Melody

5/28/80 "State cover-up at Canal alleged." NG:3-A.

Coppola, Lee
10/2/80 "Lois Gibbs makes most of meeting with President." BEN:17.

Cornwell, Martha
1980 "Organizations in a disaster setting. An adversarial model: A case study of the Love Canal chemical emergency in Niagara Falls, New York." Master's project, Department of Sociology, SUNY Buffalo.

Council on Environmental Quality
1980 "Toxic chemicals and public protection. A report to the president by the Toxic Substances Strategy Committee." Washington, D.C.: U.S. Government Printing Office.

Crotty, Peter
6/2/80 Letter to Beverly Paigen.

Cull, Jay A.
11/17/78 Letter to John Iannotti. NYDOH 1979, Love Canal [Collected Documents]: Sect. 41.

Culliton, Barbara
8/29/80 "Continuing confusion at Love Canal." *Science* 209:1002-1003.

Curts, H.J.
11/23/56 Letter to Charles B. Wright.

Dearing, Bob
2/21/79 "State Love Canal survey challenged." BCE:3.
10/27/79 "Love Canalers win major victory in battle with state." BCE:1.
5/22/81 "EPA aide raps 2 decisions on Love Canal." BCE:3.
6/16/81 "Canal tests may resume as LaFalce presses promises." BCE:3.

Delmonte, Francine
8/16/78 "Carey raises victims' hopes." NG:1-B.
8/23/78 "Profs to study Love Canal." NG:9-A.
10/8/78 "Study finds housing defects." NG:1.

Desmond, Michael
8/10/78 "Extreme caution urged in probe of Love Canal." BCE:1.
9/16/78 "Love Canal hazard is small part of national problem." BCE:1.
9/17/78 "EPA slow in fighting chemical peril." BCE:1.
9/18/78 "Illegal waste haulers create US time bombs." BCE:1.
9/19/78 "Public frets over EPA 'laxity' on dump sites." BCE:1.
9/22/78 "Love Canal tops list of disasters." BCE:1.
9/24/78 "EPA expects more suits once rules on hazardous wastes are on books." BCE:B-15.

5/11/81 Private conversation with Adeline Levine.

DeToqueville, Alexis
1956 *Democracy in America.* New York: Mentor Books (first published 1835 and 1840).

Disaster Relief Act of 1974
1974 PL 93-288; 88 U.S.C.143.

Dohrenwend, Barbara Snell
1978 "Social stress and community psychology." *American Journal of Community Psychology* 6(1):1-14.

Dowling, Edward J., Jr.
6/16/80 "Future meetings and requested materials." Memorandum to Environmental Health Panel members. Albany: N.Y. Health Planning Commission.
7/22/80 "Meeting summary of July 21, 1980 meeting; Draft report comments." Memorandum to Environmental Health Panel members. Albany, N.Y. Health Planning Commission.
12/3/80 Letter to Adeline Levine.
2/18/81 Letter to Adeline Levine.
3/25/81 Letter to Adeline Levine.

Dynes, Russell R.
1978 "Interorganizational relations in communities under stress." In E.L. Quarantelli (ed.), *Disasters: Theory and Research.* Beverly Hills, Calif.: Sage Publications.

Dynes, Russell R., and Aguirre, B.E.
1979 "Organizational adaptation to crisis: Mechanisms of coordination and structural change." *Disasters: The International Journal of Disaster Studies and Practice* 3(1):71-74.

Dynes, Russell R., and Quarantelli, Ernest L.
1970 *Organized Behavior in Disasters.* Lexington, Mass.: Lexington Books, D.C. Heath.

Ebert, Charles V.
9/21/78 "Comments on the Love Canal pollution abatement plan." Department of Geography, SUNY Buffalo.
3/79 "The physical setting of the Love Canal." Department of Geography, SUNY Buffalo.

Eckhardt, Bob
8/10/79 Letter to David Axelrod.

Economic Consultants Organization
1966 "Population: Niagara County and subdivisions." Lockport, N.Y.: Niagara County Planning Board.

Elinson, J.
1977 "Insensitive health statistics and the dilemma of the HSAs." *American Journal of Public Health* 67:417-418.

Ember, Lois
11/10/80 "Love Canal health issue still unresolved." *Chemical and Engineering News* 58:25-26.

Erikson, Kai T.
1976 *Everything in Its Path*. New York: Simon and Schuster.

Fergus, Pat
1/12/79 "Axelrod meets with canal pickets." NG:1

Flacke, Robert
Fall 1978 Draft of grant application from N.Y. Department of Environmental Conservation to USEPA. NYDOH 1979, Love Canal [Collected Documents]: Sect. 62.

Fox, Cecil H.
5/30/80 "Sakharov and whistle-blowing." *Science* 208:1976.

Francis, Mark
3/16/79 "Toxic flows abated: Hooker." NG:1-B.

Frank, Mark
10/3/80 "Carey says 'system' assisted Canal pact." BEN:3.

Freeman, Jo
1975 *The Politics of Women's Liberation*. New York: David McKay.

Freeman, Robert
1/28/81 Letter to Adeline Levine.

Friedman, William M.
5/23/78 "Public Meeting-Love Canal." Memorandum to Commissioner Berle. NYDOH 1979, Love Canal [Collected Documents]: Sect. 11.
12/1/78 Interview with Adeline Levine.

Fritz, Charles E.
1961 "Disaster." In Robert K. Merton and Robert A. Nisbet (eds.), *Contemporary Social Problems,* pp. 651-694. New York: Harcourt Brace and World.

Fuller, John G.
1977 *The Poison that Fell from the Sky*. Berkeley, Calif.: Berkeley Publishing Corporation.

Gage, Stephen J.
6/30/80 Letter to Lewis Thomas.
8/15/80 Letter to editor. *Science* 209:752.

Gibbs, Lois
8/9/79 "At Love Canal people still wait for solutions." Letter to NYT:A-20.
8/20/79 Letter to David Axelrod.
10/21/79 Letter to Joseph Maloney.
1982 *The Love Canal: My Story*. Albany, NY: SUNY Press.

Glynn, Don
2/21/81 "Environmental controversies finally took their toll on

Bruce Davis." NG:2B.

Goldsen, Rose K.

1977 *The Show and Tell Machine*. New York: Dell Publishing Company.

Governor's Office

8/3/78 Press release. Albany: State of New York Executive Chamber.

Graziano, Anthony M.

1969 "Clinical innovation and the mental health power structure: A social case history." *American Psychologist* 24(1):10-18.

Green, Sidney

5/28/80 "Review of report and slides on the pilot cytogenetic study on the residents of Love Canal, New York." Memorandum to Vilma R. Hunt, USEPA.

Greenwald, Peter

12/3/80 Letter to Adeline Levine.

Grushky, Arnold

12/24/78 Letter to William H. Wilcox. NYDOH 1979, Love Canal [Collected Documents]: Sect. 38.

Gunby, Phil

1978 "Lessons from Love Canal." *Journal of the American Medical Association* 240(9):2033.

Hadley, Wayne

8/15/78 Interview with Adeline Levine.

Hammersley, Margaret

5/20/80 "Hooker dumped 188,900 tons, firm tells court." BEN:41.

Hart, Fred D., Associates

7/28/78 "Draft report: Analysis of a groundwater contamination incident in Niagara Falls, New York." Prepared for USEPA. NYDOH, 1979, Love Canal [Collected Documents]: Sect. 31.

Haughie, Glenn E.

7/20/78 Memorandum to Drs. Axelrod, Greenwald and Vianna. NYDOH, 1979, Love Canal [Collected Documents]: Sect. 11.

3/15/79 Letter to Beverly Paigen.

Hennessy, William C.

8/6/79 "Interim report on NYS Department of Health environmental and epidemiological studies at Love Canal." Memorandum to members of the Governor's Love Canal Task Force, Albany.

Herman, Ray

10/17/79 "State OKs Love Canal home buyout." BCE:1.

5/25/80 "Federal-state quarrel over Canal may hurt Carter ballot bid." BCE:1.

10/2/80 "Happy Falls crowd hails Carter; President signs Canal aid." BCE:1.

Hildebrand, Bruce

8/14/78 "Report on Army investigation into alleged Army dumping of toxic substances in Love Canal Area, Niagara Falls, New York." Washington, D.C.: Headquarters, Department of the Army. NYDOH, 1979, Love Canal [Collected Documents]: Sect. 45.

Hitzik, Mike

8/12/78 "Canal residents get health questionnaire." BCE:2.

Hoffmann, Margeen

1980 "Progress report of the Ecumenical Task Force of the Niagara Frontier, Inc. (March 20, 1979-August 1, 1980)." Niagara Falls: Ecumenical Task Force.

Holden, Constance

6/13/80 "Love Canal residents under stress." *Science* 208:1242-1244.

Hooker Chemical Company

3/19/79 "On March 15th, Hooker held a press briefing at the Buffalo Statler Hilton. Here's a report." BEN:5 (advertisement).

5/29/79 "Try telling Bruce Davis that Hooker doesn't care about Niagara Falls." BCE:57 (advertisement).

7/13/79 "Hooker cares, and we want you to know it!" BCE:7 (advertisement).

9/5/79 "A hard look at the facts." BCE:26 (advertisement).

12/14/79 "Hooker will inherit a great deal from these employees." NG:7-B (advertisement).

1980 *Factline* #10. Niagara Falls: Hooker Public Relations Department.

10/28/80 News release. Hooker Public Affairs, Houston, Texas.

11/80 *Factline* #12. Houston: Hooker Public Affairs Department.

5/29/81 "You're about to be untricked." BEN:6 (advertisement).

Horrigan, Jeremiah

11/4/76 "PCBs come as no surprise at 99th Street." NG:3.

Hufsmith, Susan

1979 "The Response of Love Canal Residents to their Community and Government." Paper presented at the Eastern Psychological Association Meeting, Division 27, April 19.

Hunt, Vilma

1/17/81 Interview with Adeline Levine.

Janerich, Dwight T.; Burnett, William S.; Fleck, Gerald; Hoff, Margaret; Nasca, Philip; Poldenak, Anthony P.; Greenwald, Peter; and Vianna, Nicholas

6/19/81 "Cancer incidence in the Love Canal area." *Science* 212:1404-1407.

Janis, Irving L.
1972 *Groupthink*. Boston: Houghton Mifflin.

Javits, Jacob J.
7/2/80 Letter to Mr. President [Jimmy Carter].

Johnston, Scott
8/8/78 "Carey pledges $4 million 'now' in Canal aid." BEN:33.

Johnston, T.R.
7/13/78 Letter to George A. Shanahan. NYDOH, 1979, Love Canal [Collected Documents]: Sect. 12.

Kahan, Richard A.
6/4/79 Memorandum to the directors of the New York State Urban Development Corporation. Subject: Niagara Falls-Hooker Niagara Land Use Improvement Project. Albany.

Ketter, Robert
1/8/79 Interview with Adeline Levine.

Keys, Judith
12/15/78 Seminar, Love Canal research project, Department of Sociology, SUNY Buffalo.

Kilian, Jack D.
6/5/80 Letter to Vilma Hunt.

Kim, Stephen
5/22/78 "Report of the meeting on the Love Canal in Niagara Falls on Friday, May 19, 1978." Memorandum to Dr. Axelrod. NYDOH, 1979, Love Canal [Collected Documents]: Sect. 11.

Kistler, Robert
11/9/78 Letter to Arnold Grushky. NYDOH 1979, Love Canal [Collected Documents]: Sect. 38.

Klaus, Marshall H., and Kennell, John K.
1976 *Maternal-Infant Bonding*. St. Louis: C.V. Mosby.

Kolata, Gina Bari
6/13/80 "Love Canal: False alarm caused by botched study." *Science* 208:1239-1242.

Komorowski, Thad
9/26/78 "Canal rent 'mixup' reported." NG:1-B.
12/21/78 "Homeowners offered relief." NG:1.
3/10/79 "Cleanup ordered for Canal leakage." NG:1.

Kostoff, Bob
12/30/78 "Safety measures to be beefed up for dig to Canal." BCE:3.

LaFalce, John J.
6/26/78 Letter to Harold Brown. NYDOH 1979, Love Canal [Collected Documents]: Sect. 20.
12/24/78 Interview with Adeline Levine.
2/22/79 Letter to Douglas Costle and Joseph Califano.

11/9/79 Letter to President Jimmy Carter.
6/19/81 Letter to Ms. Anne Gorsuch.

Lang, F.J.
10/23/56 Letter to Acting City Manager, Niagara Falls, New York.

Lang, Kurt and Lang, Gladys
1964 "Collective responses to the threat of disaster." In George Grosser, Henry Wechsler, and Milton Greenblatt (eds.), *The Threat of Impending Disaster: Contributions to the Psychology of Stress.* Cambridge, Mass.: MIT Press.

Legator, Marvin
4/16/81 Interview with Adeline Levine.

Leonard, R.P.; Wertham, P.H.; and Ziegler, R.C.
8/77 "Characterization and abatement of groundwater pollution from Love Canal chemical land fill, Niagara Falls, New York." Calspan Corporation. NYDOH 1979, Love Canal [Collected Documents]: Sect. 4.

Lesher, Richard L.
11/3/80 "The 'Poisoning' of America?" *The Voice of Business.* Washington, D.C.: U.S. Chamber of Commerce.

Lester, Stephen U.
7/25/79 Interview with Adeline Levine.
10/25/79 Letter to David Axelrod.

Levine, Adeline
7/22/80 Letter to Donald C. Baeder.
9/9/80 Letter to Michael Tabris.
12/30/80 Letter to James McCormack.
2/19/81*a* Letter to John E. Fitzpatrick.
2/19/81*b* Letter to Nicholas Vianna.
3/5/81a Letter to Nicholas Vianna.
3/5/81*b* Letter to John E. Fitzpatrick.
3/18/81 Letter to Nicholas Vianna.
3/3/81 Letter to Charles Carter.
4/17/81 Letter to Marvin Legator (confirmation of telephone conversation).

Levine, Adeline; Barker, David; Cornwell, Martha; Hufsmith, Susan; Ploughman, Penelope; and Rowe, Sharon
1979 "The Love Canal: A Sociologist's Perspective." Paper presented at the Eastern Sociological Society Meeting, March 16.

Levine, Adeline, and Levine, Murray
1975 "Evaluation research in mental health: Lessons from history." In J. Zusman and C. Wurster (eds.), *Program Evaluation: Alcohol, Drug Abuse, and Mental Health Service Programs.* Lexington, Mass.: Lexington Books, D.C. Heath.

1977 "The social context of evaluative research: A case study." *Evaluation Quarterly* 1(4):515-542.

Levine, Murray
1/2/81 Letter to editor. *Science* 211:8.

Logan, Andy
7/30/79 Around City Hall: Same time, next year." *New Yorker*, pp. 78-82.

Love Canal Homeowners Association
1980 "Love Canal Chronological Report, April 1978 to January, 1980." Niagara Falls.
6/16/80 Press release.

Love Canal Task Force Temporary Relocation Program
9/79 Physician's statement.

Lynch, David E.
8/5/78 "$4 million Love Canal cleanup OK'd: Senate due to approve on Monday." BCE:1.
10/14/79 "House unit urges Canal evacuation." BCE:1.
5/25/80 "Love Canal study puts feds in a frenzy." BCE:C-1.
6/8/80 " 'Superfund' gets push ahead at Senate hearing on waste." BCE:1.
11/25/80 "Senate cuts, OKs Superfund bill." BCE:3.

Lynch, David E., and Moe, Kristine
5/18/80 "790 Canal families may evacuate." BCE:1.

McCarthy, John D., and Zald, Mayer M.
1977 "Resource mobilization and social movements: A partial theory." *American Journal of Sociology* 82(6):1212-1241.

McCarthy, Max
8/11/78 "Canal crisis shows EPA it lacks power, funds." BEN:3.

MacClennan, Paul
8/3/78 "Cancer researcher urges relocation of 50 families near toxic dump site." BEN:13.
12/17/78 "State plans new aid package to defuse Love Canal protest." BEN:C-4.
12/20/78 "Hooker dump report raises fears of dioxin contamination." BEN:1.
1/5/79 "State extends probes north of Canal." BEN:1.
1/29/79 "Canal residents plan push for full evacuation." BEN:13.
2/20/79 "More Canal evacuations are urged: Researcher warns state of health risks." BEN:1.
5/11/79 "Canal runoff draining into Niagara River." BEN:1.
10/18/79 "Carey will act to buy Canal homes: State purchase of 239 sites being planned." BEN:1.
5/18/80 "U.S. prepares to evacuate all Love Canal families." BEN:1.

5/19/80 "White House blocked Canal pullout." BEN:1.
5/20/80*a* "EPA recalls two officials held by crowd." BEN:39.
5/20/80*b* "Canal study review stirs dispute: Research firm eyes naming rival panel." BEN:1.
5/25/80 "Canal studies bypass area, experts claim." BEN:6.
6/12/80 "State failed to install wells to monitor Canal." BEN:37.
6/18/80 "Cuban refugee costs may have sidetracked U.S. plans for Canal." BEN:1.
8/1/80 "State likely to OK Canal loan plan: U.S. offers $15 million to buy homes." BEN:1.
8/5/80 "State rejects Canal loan. Seeks changes. BEN:1.
8/21/80 "Love Canal loan talks break down: $15 million is snagged in impasse." BEN:1.
8/23/80 "Pact reached on Canal relocation: U.S. plans to provide $15 million." BEN:1.
10/26/80 "Carey's panel fails to find root of Love Canal fiasco." BEN:E-6.
11/5/80 "EPA finds Canal wastes decreasing." BEN:39-II.
2/12/81 "Coast Guard kills funds to clean up dioxin in sewers." BEN:33.
3/14/81 "Testing at Love Canal shelved." BEN:1.
6/20/81 "Agency to move cautiously on selling Canal homes, fearing future suits." BEN:B-3.
7/24/81 "Paigens resigning posts at Roswell Institute." BEN:6.

MacClennan, Paul, and Shribman, David
8/11/78 "Canal remedies may be too limited." BEN:1.
8/13/78 "Legal action on Canal site urged in '77" BEN:1.
8/22/78 "$7.5 million price put on Canal homes." BEN:33.
8/24/78 "Carey again defuses Love Canal discord." BEN:15.
8/26/78 "State, U.S. are fumbling over Love Canal tab." BEN:B-2.
10/12/78 "State rejects requests of families outside Canal perimeter." BEN:19.
11/16/78 "U.S. Canal aid draws ire of state officials." BEN:39.
5/22/80 "Canal home-buy faces delay: Test results called key to final decision." BEN:1.
5/23/80 "Canal to cost another $80 million: But dispute looms over State-U.S. split." BEN:1.

McCormack, James
12/29/80 Private communication with Adeline Levine.

MacDonald, Dan
9/20/78 "State tests show PCBs in water, soil from Canal." BEN:1.
4/28/80 "State asks $635 million in suit against Hooker." BEN:25.
5/20/80 "EPA insists incident won't affect Love Canal decision." BEN:39.

8/12/80 "Canal group will picket convention." BEN:24.
4/18/81 "EPA stalling Canal study, residents say." BEN:B-3

McMahon, John
8/9/78 Memorandum on Love Canal to Robert L. Collin, NYDEC, Albany.

McNeil, Donald G.
8/2/78 "Upstate waste site may endanger lives." NYT:1.
8/8/78 "Carter approves emergency help for Love Canal." NYT: 1.

Maloney, Joseph
6/12/79 Letter to Mary Hogan.
11/9/79 Letter to Lois Gibbs.
7/7/81 Interview with Adeline Levine.

Mechanic, D., and Newton, M.
1965 "Some problems in the analysis of morbidity data." *Journal of Chronic Diseases* 18:560-580.

Meislin, Richard L.
5/25/80 "Love Canal residents say the state has failed them." NYT:E-6.
6/9/80 "Question raised by call for new Love Canal study." NYT:B-4.
10/11/80 "Panel discounts 2 studies on Love Canal problems." NYT:25.

Merton, Robert K.
1973 *The Sociology of Science*. Chicago: University of Chicago Press.

Merton, Robert K., and Kitt, Alice S.
1950 "Contributions to the theory of reference group behavior." In Robert K. Merton and Paul Lazarsfeld (eds.), *Continuities in Social Research*. Glencoe, Ill.: Free Press.

Millock, Peter
1979 Draft report, Interagency Task Force on Hazardous Wastes. Albany.

Moe, Kristine
5/17/80 "Genetic harm tied to Canal." BCE:1.
10/11/80 "Canal health studies hit by doctors' panel." BCE:1.
12/14/80 "Researchers rap state's review of health studies at Love Canal." BCE:A-2.

Molotch, Harvey, and Lester, Marilyn
1975 "Accidental news: The great oil spill as social occurrence and national event." *American Journal of Sociology* 81(2):235-260.

Molotsky, Irvin

12/21/79 "Hooker is sued by Justice Department over Love Canal." NYT:B-2.
5/17/80 "Chromosome damage found in Love Canal tests." NYT:1.
5/18/80 "710 more families in Love Canal area may be relocated. NYT:1.
8/23/80 "U.S. agrees to $7.5 million for Love Canal residents, along with loans." NYT:27.

Morgado, Robert J.
2/21/79 Letter to Jack Watson, Jr. NYDOH 1979, Love Canal [Collected Documents]: Sect. 38.

Moriarty, Lawrence R.
10/18/77 Memorandum to William LiBrizzi. NYDOH 1979, Love Canal [Collected Documents]: Sect. 49.

Murphy, Gerald
8/6/78 Memorandum to Dr. Mirand. Sect. 6 in Beverly Paigen, *Notebook*, 3/7/80.
12/4/78 Letter to Dr. Mirand, Dr. Pressman, cc Mr. Delellis. Sect. 3 in Beverly Paigen, *Notebook*, 3/7/80.

National Institute of Environmental Health Sciences
7/80 *Federal Agency Support for Environmental Health Research*. Research Triangle Park, N.C.: NIEHS.

Nelkin, Dorothy (ed.)
1979 *Controversy: Politics of Technical Decisions*. Beverly Hills, Calif.: Sage Publications.

New York Department of Health (NYDOH)
9/78 "Love Canal: Public health time bomb." A special report to the governor and legislature. Albany.
12/11/78 New release. Albany: Health Communications.
1979 Love Canal [Collected Documents.] Albany.
6/23/80*a* "Fact sheet. State Health Department's role at Love Canal." Albany: Health Communications.
6/28/80*b* News release, "Note to editors and reporters." Albany: Health Communciations.
4/81 "Love Canal: A special report to the governor and legislature." Albany.

New York Department of State
1979 *Manual for the Use of the Legislature of the State of New York, 1977-1979*. Albany.

New York Department of Transportation (NYDOT)
11/78 Application to the FDAA. NYDOH 1979, Love Canal [Collected Documents]: Sect. 38.
1979 Grant agreement with United Way of Niagara. NYDOH 1979, Love Canal [Collected Documents]: Sect. 40.

9/15/78 Love Canal Task Force, information bulletin.

12/31/80 "Actual/Projected Costs Based on Commitments through 12/31/80 and Estimated Federal Participation." Albany, New York.

New York Executive Law (McKinney's)

1979 Section 6

New York Laws

1979 Ch. 732, 1979, N.Y. Laws, 1.

1980 Ch. 259, 1980, N.Y. Laws. 404 (McKinney's) Sect. 2. Codified as N.Y. General Municipal Law, Sect. 950.

New York Public Health Law (McKinney's)

1979 Sections 206; 1389; 2801:5; 2802

New York Public Officer's Law (McKinney's)

1979 Articles 6 and 7. Sections 74; 100; 101

New York Real Property Tax Law (McKinney's)

1979 Laws of New York, 1979. Ch. 2. Sect. 1; Ch. 258, Sect. 1; Ch. 703, Sect. 1.

New York State Health Planning Commission

1980 "Chronological summary: Love Canal, 1940's-1979 and Love Canal chronology." Albany.

New York Times (NYT)

12/17/78 "Environmental toxology (sic) expert appointed health chief by Carey." NYT:45

1/22/80 "Regan approves loan to Hooker Chemical despite doubts over 'social responsibility.' " NYT:B-1.

5/21/80 "Agreement reached to move out some of 710 Love Canal families." NYT:1.

6/12/80 "Gobbledygook bomb at Love Canal." NYT:30-A (editorial).

10/17/80 "Those disastrous studies at Love Canal." NYT:30-A (editorial).

6/14/81 "No final verdict on Love Canal cancers." NYT:6-E.

Niagara Gazette (NG)

10/17/53 "Hooker gives land for school and park in 97th Street area." NG:11 (editorial).

11/24/59 "City to acquire tract for recreation area." NG:1.

7/14/60 "City to use land at 99th St. School for play purposes." NG:4.

11/4/76 "Lots near school are put off limits." NG:1-B.

8/15/77 "Seal off Love Canal." NG:10-A (editorial).

9/12/77 "Councilman dumps on LaFalce's' tour.' " NG:1-B.

4/28/78 "State orders Love Canal site cleanup." NG:1.

8/5/78*a* "State funds go for evacuation." NG:1.

8/5/78*b* "Krupsak criticizes Carey on Love Canal." NG:2.

8/6/78 "Gazette continues Canal role." NG:1.
8/12/78 "United Way takes helm of Canal relief programs." NG:1.
8/13/78 "Canal aid offers continue." NG:4-B.
8/14/78 "Canal safety rules upset Homeowners." NG:1.
10/6/78 "3 Canal area families relocations extended by additional funds." NG:1-B.
3/14/79 "Religious leaders join Canal effort." NG:1.
3/23/79*a* "Axelrod tells of kidnap threat." NG:1.
3/23/79*b* "Researcher claims 'conflict' in Canal dealings." NG:1-B.
7/15/79 "Churches' Canal Task Force hires director." NG:1-B.
10/4/79 "Fonda joins relocation fight." NG:1.
10/11/79 "Not an answer." NG:4-A (editorial).
5/17/80 "EPA to release new Canal health findings today." NG:1.
11/17/80 "Lack of accountants delays Canal probe." NG:3-A.
4/22/81 "Canal residents victims of presidential transition." NG: 4-A (editorial).
5/18/81 "Morris may purchase home in Love Canal." NG:5-A.
5/27/81 "Morris' proposed move to Canal home is premature." NG:6-A (editorial)

Occidental Petroleum Corporation
1978 "Annual Report for the Year 1978."

O'Laughlin, Michael
10/24/80 Interview with Adeline Levine.

Olsen, Marvin
1968 *The Process of Social Organization*. New York: Holt, Rinehart and Winston.

Omicinski, John
5/22/80 "Carey urges lasting relocation of Canalers: Temporary federal solutions hit." NG:1.

Osborn, John E.
1977 "New York's Urban Development Corporation. A study of the unchecked power of a public authority." *Brooklyn Law Review* 43(2):237-282.

Packard, Vernon L.
10/19/61 Letter to Frank J. Lang.

Page, Arthur
5/22/80 "Thorough study of Canal families will begin soon." BEN:19.
6/4/80 "Board discounts state censorship at Roswell Park." BEN:23-II.

Page, Arthur, and Shribman, David
6/26/80 "State delayed releasing Canal study to panels." BEN:17.

Paigen, Beverly
10/31/78 "Preliminary analysis of health effects." Memorandum to

Love Canal Homeowners Association. Buffalo: Roswell Park Memorial Institute.

11/28/78 Letter to Robert Whalen. Sect. 3 in Beverly Paigen, *Notebook*, 3/7/80.

12/19/78 "Miscarriages in Love Canal residents." Memorandum to Lois Gibbs and Elena Thornton. Buffalo: Roswell Park Memorial Institute.

1/8/79 "Release of data." Memorandum to Glenn Haughie. Buffalo: Roswell Park Memorial Institute.

2/16/79 "Analysis of birth defects, miscarriages and birth weights over time." Memorandum to Nicholas Vianna.

3/21/79 "Health hazards at Love Canal." Testimony presented to the House Subcommittee on Oversight and Investigation.

3/29/79 "Methodological Problems with the Love Canal Health Studies Conducted by The State of New York." Memorandum to David Rall.

5/79 Interview with Adeline Levine.

9/19/79 Letter to David Axelrod. Sect. 5 in Beverly Paigen, *Notebook*, 3/7/80.

12/21/79 Letter to David Axelrod.

3/7/80 *Notebook*. Prepared for Roswell Park Memorial Institute Board of Visitors. Buffalo.

6/12/80 Letter to Peter Crotty.

6/18/80 Letter to Robert J. Freeman.

1/2/81 Letter to editor. *Science* 211:8.

Palazzetti, Agnes

8/27/78 " '54 Alarm' ignored on Canal site." BEN:1.

Parry, David (with the assistance of Kevin Ferry, Walter P. Nagely and Marty Sienkiewicz)

1975 "William T. Love and the development of Model City." Unpublished manuscript, School of Architecture and Environmental Design, SUNY Buffalo.

Peck, Louis

5/22/80 "Politics at work even in trauma of the Love Canal." NG:8-A.

7/3/80 "Congress, White House at odds over intent of Love Canal relocation funding." NG:4-A.

2/11/81 "Cut Niagara pollution Canada, U.S. told." NG:1.

6/16/81 "3 N.Y. legislators ask for Love Canal health study." NG:1.

7/9/81 "Proposed Canal health study will be evaluated next month." NG:1.

Penca, Jack

6/20/80 "US reply to Dr. Axelrod's memorandum in opposition to motion to compel subpoena compliance." Item #51, *U.S.* v *Hooker*, Civ. No. 79-99. (W.D. N.Y. 1979).

Perrault, Larry
8/8/78 "State to assume Canal mortgages." NG:1.
8/18/78 "Canal repair plan criticized by residents." NG:1.
8/24/78 "State will review final Canal plans." NG:1.
8/30/78 "Canal forum becomes bedlam." NG:1-B.

Picciano, Dante
1980 "A pilot cytogenetic study of the residents living near Love Canal, a hazardous waste site." *Mammalian Chromosome Newsletter* 21(3).
5/5/80 Letter to Frode Ulvedal. In USHR: 5/22/80:22.
8/15/80 Letter to editor. *Science* 209:754-756.

Ploughman, Penelope
1979 "Mass Media in Disaster: Love Canal Coverage." Paper presented at the Eastern Psychological Association Meetings, Division 27, April 19.

Pollak, David
10/3/76 "Closeup: Hooker dump troubles neighbors in LaSalle." NG:1.
5/1/77 "Dump seepage tested; City needs plan." NG:1.
5/3/77 "Seepage health threat slight." NG:1-B.
6/5/77 "Family recalls fires at Hooker dump." NG:1.

Porter, Sabrina
5/17/79 "Canal engineers stir ire." NG:1-B.
5/20/79 "Love Canal homes up for grabs." NG:1-B.
8/22/79 "Canal meeting offers little solace." NG:1.

Powell, Roland
3/23/79 "Canal illnesses hard to verify, Axelrod testifies." BEN:7.
7/26/79 "U.S. widens health concern in Canal area." BEN:1.
9/6/79 "Persistence is her watchword as Mrs. Gibbs goes to Washington." BEN:9.
6/14/80 "Javits, Moynihan to press Congress on relocations." BEN:3-B.

President's Commission on Mental Health
1978 *Report to the President*, Vol. I. Washington, D.C.:U.S. Government Printing Office.

Pressman, David
9/7/78 Memorandum to Dr. Beverly Paigen. "The Proposal Entitled 'A Proposal for Aerosol Formation Studies Involving SO_2 and Diesel Exhaust in Smog Chambers, Calspan Proposal #6038' prepared for the Environmental Protection Agency." Memorandum to Dr. Beverly Paigen. Sect. 8 in Beverly Paigen, *Notebook*, 3/7/80.

Quarantelli, Ernest, and Dynes, Russell R.
1977 "Response to social crisis and disaster." *Annual Review of Sociology*.

Rall, David
7/26/79 "Report of meeting between scientists from HEW and EPA and Dr. Beverly Paigen and scientists of the NYDOH concerning Love Canal." Research Triangle Park, N.C.: NIEHS.
6/30/80 Letter to Lewis Thomas.
12/29/80 Letter to Adeline Levine.

Rosen, Jay
8/6/78 "Lawyers eye landmark Love Canal path to court." BCE: B-1.

Rowe, Sharon Kay (Masters)
1979 "Victim Life Stage Perceptions of Environmental Crisis." Paper presented at the Eastern Psychological Association Meetings, Division 27, April 19.

Russell, David L.
11/6/76 "State sets probe at chemical dump." NG:B-1.
11/9/76 "State will sample drainage from former Hooker dump." NG:B-1.

Russell, David L., and Pollak, David
11/2/76 "Dangerous chemicals found leaking from Hooker dump." NG:1.
11/4/76 "Toxic chemicals carried in storm sewers." NG:B-1.

Safranek, Ed
8/6/78 "Canal area homeowners split; Seek quick aid." NG:1.
7/6/79 "Canal Homeowners caught in quandary." NG:B-1.

Sanders, Joyce
5/1/79 "A curse just for whites." NG:A-4 (letter to editor).

Sarason, Seymour B.
1972 *The Creation of Settings and the Future Societies.* San Francisco: Jossey-Bass.

Schoenfeld, A. Clay; Meier, Robert F; and Griffin, Robert J.
10/79 "Constructing a social problem: The press and the environment." *Social Problems* 27:38-61.

Seal, Geoff
7/28/78 "Falls halts Love Canal work funds." BCE:17.
8/3/78 "Health department asks part evacuation in Love site." BCE:1.
8/6/78 "Disaster chief sure of Love Canal aid: Tour convinces official of need; Carter report due." BCE:1.

Sercu, Charles L.
1/7/81 "Hazardous waste management. Some key issues. An industrial viewpoint." Paper presented at AAAS meeting, Toronto.

Severo, Richard
5/27/80 "A tangle of science and politics lies behind study at Love

Canal." NYT:1.

Shaw, Margery W.

8/15/80 "Love Canal chromosome study." Letter to editor, *Science* 209:751-752.

Shribman, David

8/9/78 "State to buy homes bordering Love Canal." BEN:1.

8/27/78 "Engineers optimistic on Canal." BEN:B-1.

9/26/78 "State agrees to new tests at Canal." BEN:41.

10/8/78 "Renters ask halt in Canal plan." BEN:B-1.

10/9/78 "Fight looms over $175,000 in lost funds." BEN:1.

10/10/78 "Love Canal cleanup goes into full swing as yards are cleared for drainage." BEN:41.

11/12/78 "Canal probers fear effects of dioxin." BEN:B-4.

12/12/78 "8 more Love Canal pickets arrested at cleanup site." BEN:1.

2/15/79 "Canal Homeowners take coffin to Carey's office." BEN:1.

2/22/79 "Carey rejects appeal to buy more homes." BEN:1.

8/19/79 "Love Canal rites are haunted by painful memories." BEN:B-1.

8/21/79 "Dioxin is confirmed at Love Canal: State finding stirs pleas of Homeowners." BEN:1.

10/14/79 "House study asks Canal evacuation." BEN:1.

5/22/80 "EPA doubted genetic study House is told." BEN:1.

5/25/80 "How Love Canal reached White House." BEN:E-4.

Shribman, David, and Johnston, Scott

2/9/79 "More families to be moved at Love Canal." BEN:1.

Shribman, David, and MacClennan, Paul

8/12/78 "Safety to be paramount in Love Canal cleanup." BEN:1.

8/14/78 "Love Canal chemicals make 'hazardous list.' " BEN:1.

8/16/78 "Canal residents calmer as fear yields to reason." BEN: 35.

8/23/78 "Love Canal costs put at $22 million: Chances slim U.S. will offer much in aid." BEN:1.

8/25/78 "New evacuations at Love Canal to be limited." BEN:3.

8/29/78 "Project families feel 'ignored' in Canal plan." BEN:1.

9/1/78 "Officials chart legal moves in Canal dispute." BEN:1.

10/18/78 "Wider patterns of Canal illness found: Stream beds trace paths of sickness." BEN:1.

10/28/78 "President will face Love Canal protest." BEN:1.

11/1/78 "Data shows illness high near Canal group claims." BEN:26.

5/26/79 "Sale of Love Canal homes stirs protests, few bids."

BEN:1.
5/29/79 "Canal group questioned on new tactics." BEN:37.
5/28/80 "Canal impasse forcing issue to Congress." BEN:33.
5/31/80 "U.S. refuses Love Canal relocation: Final move state's, job aide insists." BEN:1.
6/5/80 "U.S. refusal angers Carey, Gibbs: Cannot buy Canal homes, U.S. aide says." BEN:1.

Silver, Bob
3/5/79 "Love Canal 'killing church.' " NG:B-1.
4/13/79 "County, city limit Love Canal talk." NG:B-1.
9/2/79 "Legislators send Love Canal plea: Relocation requested for Love Canal families." NG:1.

Sitton, Paul L.
1/30/81 Letter to Adeline Levine.

Smith, R. Jeffrey
10/3/80 "Love Canal reviewed." *Science* 210:513.
1/2/81 Letter to editor. *Science* 211:8.

Solovitch, Sara
5/26/80 "Canalers await medical tests." BCE:34.

Spencer, Gary
8/17/78 "City-State discord arises over Canal." BCE:2.
8/19/78 "Love Canal poses question: What is an emergency?" BCE:4.
11/11/78 "Canal force warned about deadly chemicals." BCE:3.

Stallings, Robert A.
1978 "The structural patterns of four types of organizations in disaster." In E.L. Quarantelli (ed.), *Disasters: Theory and Research*, pp. 87-103. Beverly Hills, Calif.: Sage Publications.

State Federal Information Bulletin
5/26/80 "Relocation information federal emergency—Love Canal area." Albany: NYDOT.

Stevens, Carol
5/25/80 "Love Canal researcher details harassment charge." BCE:1.

Stone, Christopher
1975 *Where the Law Ends*. New York: Random House.

Stutz, Eric
7/1/78 "3rd Love Canal study slated." NG:B-4.
9/7/78 "LaSalle development residents hit Task Force." NG:B-4.

Swan, Jon
1-2/79 "Uncovering Love Canal." *Columbia Journalism Review*, pp. 46-51.

Tarlton, Frances
10/7/80 Letter to Adeline Levine.

Thiele, Charles I.
1/21/54 Letter to Wesley L. Kester.
1/29/54 Letter to Wesley L. Kester.
6/25/56 Letter to P. Friona.
10/15/56 "Report on site development work, 99th Street Elementary School. Niagara Falls, New York, Board of Education.
10/18/56 Letter to Arthur Silberberg.

Thomas Lewis
7/3/80 Letter to Lois Gibbs.
10/8/80 Letter to The Honorable Hugh L. Carey and Members of the New York State Legislature.
10/80 "Report of the Governor's Panel to Review Scientific Studies and the Development of Public Policy on Problems Resulting from Hazardous Wastes."
11/4/80 Letter to Adeline Levine.

Thomas, Robert E.
1955 *Salt & Water, Power & People*. Niagara Falls: Hooker Electrochemical Company.

Time Magazine
11/3/80 "Another look at Love Canal." *Time*, p. 99.

Tsubaki, Tadao, and Iru, Katsura
1977 *Minamata Disease*. New York: Elsevier.

Tully, James H., Jr.
7/11/80 Letter to Beverly Paigen.

Tyson, Rae
5/27/80 "Love Canal groups to boycott EPA health tests." NG:3.
6/4/80 "State: Canal relocation rate 50%." NG:3.
6/18/80 "EPA promises Canal safety report by December." NG:3.
10/29/80 "Suit calls data unsupported." NG:3.
5/7/81 "Wesley Methodist to close, a victim of the Love Canal." NG:3.
5/19/81 "Task Force questions Morris' plan of moving into vacant Canal home." NG:3.
6/4/81 "2 homes show signs of chemical migration." NG:1.
7/8/81 "Delay seen in release of Canal tests." NG:1.

U/B Alumni News
9/78 13(3):1, U/B Alumni Association, SUNY Buffalo.

U.S. v *Hooker*
12/20/79 Civ. No. 79-990 (W.D.N.Y. 1979).

U.S. Congress
1911 "Treaty between the United States and Great Britain

relating to boundary waters between the United States and Canada." *The Statutes at Large of the United States of America from March 1909-March 1911.* Vol. XXXVI, Part 2:2448-2455.
U.S. Congress, 61st Sess., 1909-1911.
7/3/80 Public Law 96-304.

U.S. Department of Commerce
10/3/50 *1950 Census of the Population. Preliminary Counts.* Bureau of the Census. Series PC-2#47.
5/78 *Survey of Current Business* 58(5):S-1, S-6, 56.

U.S. Department of Health, Education and Welfare (USDHEW)
7/26/79 Press release. "Secretary Joseph A. Califano made public today the findings of a panel of scientists . . . "
2/80 *Report of the Subcommittee on the Potential Health Effects of Toxic Chemical Dumps of the DHEW Committee to Coordinate Environmental and Related Programs.* Research Triangle Park, N.C.: NIEHS.

U.S. Environmental Protection Agency (USEPA)
1980 "Everybody's Problem: Hazardous Wastes." SW-826. Washington, D.C.: U.S. Government Printing Office.
2/21/80 "Clean Water Act 311-Actions Concerning Love Canal." *EPA Environmental Facts.*
5/17/80 Press release. "EPA finds chromosome damage at Love Canal."
5/21/80 "EPA, New York State announce temporary relocation of Love Canal residents." *Environmental News.*

U.S. Department of Health and Human Services (USDHHS)
1980 "Assessment of the threat to public health posed by toxic chemicals in the United States." A report to the United States Senate Committee on Environment and Public Works Subcommittee on Environmental Pollution.

U.S. House of Representatives (USHR)
3/21/79 et seq. "Part I. Hazardous waste disposal hearings before the Subcommittee on Oversight and Investigations of the Committee on Interstate and Foreign Commerce." 96th Congress, 1st Session, Serial #96-48. Washington, D.C.: U.S. Government Printing Office.
5/22/80 "Love Canal: Health studies and relocation." Joint hearing before the Subcommittee on Oversight and Investigations of the Committee on Interstate and Foreign Commerce, and the Subcommittee on Environment, Energy and Resources of the Committee on Government Operations. 96th Congress, 2nd Session, Serial #96-191. Washington, D.C.: U.S. Government Printing Office.

9/79 "Hazardous Waste Disposal." Report together with additional and separate views by the Subcommittee on Oversight and Investigations of the Committee on Interstate and Foreign Commerce. 96th Congress, 1st Session. Washington, D.C.: U.S. Government Printing Office.

U.S. Senate

5/18/79 et seq. "Hazardous and Toxic Waste Disposal Field Hearings." Joint hearings before the Subcommittee on Environmental Pollution and Resource Protection of the Committee on Environment and Public Works. Part 2, Serial #96-H9. 96th Congress, 1st Session. Washington, D.C.: U.S. Government Printing Office.

6/6/80 "Health Effects of Hazardous Waste Disposal Practices, 1980." Joint hearings before the Subcommittee on Health and Scientific Research of the Committee on Labor and Human Resources and the Committee on the Judiciary. 96th Congress, 2nd Session. Washington, D.C.: U.S. Government Printing Office.

United Way of Niagara

9/79 "United Way of Niagara Inc. Love Canal Account."

Vianna, Nicholas

10/78 Letter to Dear (See Attached List). NYDOH 1979, Love Canal [Collected Documents]: Sect. 58.

6/19/80 "Love Canal investigations conducted by the Bureau of Environmental Epidemiology and Occupational Health." Memorandum to Dr. Greenwald. Albany: NYDOH.

3/12/81 Letter to Adeline Levine.

3/24/81 Letter to Adeline Levine.

6/25/81 Letter to Adeline Levine.

Vianna, Nicholas J.; Polan, Adele K.; Regal, Ronald; Kim, Stephen; Haughie, Glenn E.; and Mitchell, Douglas

4/80 "Adverse Pregnancy Outcomes in the Love Canal Area." (Provisional.) Albany: NYDOH.

Violanti, J.

5/24/78 "Love Canal—Public Meeting, May 19, 1978." Memorandum to Dr. Campbell. NYDOH 1979, Love Canal [Collected Documents]: Sect. 11.

Vogel, Mike

10/4/79 "Fonda Canal protest debuts at City Hall." BEN:15.

Wall Street Journal

10/20/80 "Toxic Science." Editorial, p. 26.

Warburton, Dorothy, and Fraser, F. Clarke

3/64 "Spontaneous abortion risks in man: Data from reproductive histories collected in a medical genetics unit. *American Journal of Human Genetics* 16:1-25.

Warner, Alice
7/26/78 "Burned off paint." NG:10 (letter to editor).

Warner, Gene
4/1/81 "Magazine sued over criticism of Love Canal study." BEN:24.

Wells, P.T.
8/18/78 "Love Canal Task Force Meeting." Memorandum to M.J. Cuddy. NYDOH 1979, Love Canal [Collected Documents]: Sect. 37.

Westmoore, Paul
8/9/78 "School lawyer's Canal site advice 'unheeded' " NG:B-1.
11/18/78 "Hartford to sue on Canal." NG:1.
7/1/80 "Love Canal agency eyes appointment." NG:3.
10/18/80 "Love Canal cleanup probe delays federal funds to city." NG:1.

Whalen, Robert P.
4/25/78 Letter to Francis J. Clifford.
6/20/78 "In the matter of the Love Canal chemical waste landfill site located in the City of Niagara Falls, Niagara County, State of New York." Order.
8/1/78 "Love Canal—Current Status and Recommendations." Memorandum to Thomas Frey. NYDOH 1979, Love Canal [Collected Documents]: Sect. 27.
8/2/78*a* "In the matter of the Love Canal chemical waste landfill site located in the City of Niagara Falls, Niagara County." Order. NYDOH 1979, Love Canal [Collected Documents]: Sect. 28.
8/2/78*b* "In the matter of the Love Canal chemical waste landfill site located in the City of Niagara Falls, Niagara County." Order. Pp. 27-32. In NYDOH 9/78, "Love Canal: Public Health Time Bomb."
11/28/78 Letter to Gerald Murphy. In Sect. 13 of Beverly Paigen, *Notebook*, 3/7/80.

Whiteside, Thomas
9/4/77 "Contaminated." *New Yorker*, pp. 34-81.
1979 *The Pendulum and the Toxic Cloud: The Dioxin Threat from Vietnam to Seveso*. New Haven: Yale University Press.

Wilcox, Ansley
11/21/57 Letter to Charles M. Bent.

Wilcox, William

1/19/79 Letter to Arnold Grushky. NYDOH 1979, Love Canal [Collected Documents]: Sect. 38.

Williamson, Shelly
1979 "Technical review of the New York State Department of Health proposal for epidemiology studies." Memorandum to Stephen J. Gage. Washington, D.C.: USEPA.

Wolfe, Tom
1970 *Radical Chic and Maumauing the Flak Catchers*. New York: Farrar, Straus and Giroux.

World Health Organization
6/78 "Long-term hazards of polychlorinated dibenzodioxins and polychlorinated dibenzofurans." Joint report of U.S. National Institute of Environmental Health Sciences/ International Agency of Research on Cancer *ad hoc* working group.

Zeusse, Eric
2/81 "Love Canal: The truth seeps out." *Reason*, pp. 17-33.

Index

About the Author

Adeline Gordon Levine is an associate professor of sociology at the State University of New York at Buffalo. She received the B.A. in 1962 from Beaver College, and the Ph.D. in sociology from Yale University in 1968. Before that, she earned a nursing degree and worked as a registered nurse for five years. Her publications include *A Social History of Helping Services* (with Murray Levine, 1970), articles on aspects of women's careers, and sociohistorical factors in evaluation. This book is the culmination of three years of work and observation at Love Canal.